MATHEMATICS AND 21ST CENTURY BIOLOGY

Committee on Mathematical Sciences Research for
DOE's Computational Biology

Board on Mathematical Sciences and Their Applications

Division on Engineering and Physical Sciences

NATIONAL RESEARCH COUNCIL
OF THE NATIONAL ACADEMIES

D1445796

THE NATIONAL ACADEMIES PRESS
Washington, D.C.
www.nap.edu

THE NATIONAL ACADEMIES PRESS   500 Fifth Street, N.W.   Washington, DC 20001

NOTICE: The project that is the subject of this report was approved by the Governing Board of the National Research Council, whose members are drawn from the councils of the National Academy of Sciences, the National Academy of Engineering, and the Institute of Medicine. The members of the committee responsible for the report were chosen for their special competences and with regard for appropriate balance.

This study was supported by Contract No. DE-AT01-03ER25552 between the National Academy of Sciences and the Department of Energy. Any opinions, findings, conclusions, or recommendations expressed in this publication are those of the author(s) and do not necessarily reflect the views of the organizations or agencies that provided support for the project.

International Standard Book Number 0-309-09584-0 (Book)
International Standard Book Number 0-309-54856-X (PDF)

Library of Congress Catalog Card Number 2005024164

Additional copies of this report are available from the National Academies Press, 500 Fifth Street, N.W., Lockbox 285, Washington, DC 20055; (800) 624-6242 or (202) 334-3313 (in the Washington metropolitan area); Internet, http://www.nap.edu

THE NATIONAL ACADEMIES
Advisers to the Nation on Science, Engineering, and Medicine

The **National Academy of Sciences** is a private, nonprofit, self-perpetuating society of distinguished scholars engaged in scientific and engineering research, dedicated to the furtherance of science and technology and to their use for the general welfare. Upon the authority of the charter granted to it by the Congress in 1863, the Academy has a mandate that requires it to advise the federal government on scientific and technical matters. Dr. Bruce M. Alberts is president of the National Academy of Sciences.

The **National Academy of Engineering** was established in 1964, under the charter of the National Academy of Sciences, as a parallel organization of outstanding engineers. It is autonomous in its administration and in the selection of its members, sharing with the National Academy of Sciences the responsibility for advising the federal government. The National Academy of Engineering also sponsors engineering programs aimed at meeting national needs, encourages education and research, and recognizes the superior achievements of engineers. Dr. Wm. A. Wulf is president of the National Academy of Engineering.

The **Institute of Medicine** was established in 1970 by the National Academy of Sciences to secure the services of eminent members of appropriate professions in the examination of policy matters pertaining to the health of the public. The Institute acts under the responsibility given to the National Academy of Sciences by its congressional charter to be an adviser to the federal government and, upon its own initiative, to identify issues of medical care, research, and education. Dr. Harvey V. Fineberg is president of the Institute of Medicine.

The **National Research Council** was organized by the National Academy of Sciences in 1916 to associate the broad community of science and technology with the Academy's purposes of furthering knowledge and advising the federal government. Functioning in accordance with general policies determined by the Academy, the Council has become the principal operating agency of both the National Academy of Sciences and the National Academy of Engineering in providing services to the government, the public, and the scientific and engineering communities. The Council is administered jointly by both Academies and the Institute of Medicine. Dr. Bruce M. Alberts and Dr. Wm. A. Wulf are chair and vice chair, respectively, of the National Research Council.

www.national-academies.org

Preface

This report was commissioned by the Office of Advanced Scientific Computing Research (OASCR) at the Department of Energy (DOE). This office, which has broad responsibilities for applications of mathematics and computing to all fields of science of importance to DOE, sought advice as specified in the charge to the committee:

> The study will recommend mathematical sciences research activities to the Department of Energy that will enable science to make effective use of the large amount of existing genomic information and the much larger and more diverse collections of structural and functional genomic information that are being created. The recommended activities should cover both current research needs and also include some higher-risk research that might lead to innovative approaches for the future.

In discussions with OASCR officials, it became apparent that the intent was to sponsor a broad, scientifically based view of the opportunities that now lie at the interface between the mathematical sciences and biology. "The mathematical sciences" was to be broadly defined to include statistics, computational science, and all areas of applied mathematics.[1] Although the Department of Energy is an agency with deep roots in applying the mathematical sciences to the physical sciences—as well as a pioneer in selected biological applications such as protein-structure de-

---

[1]An upcoming National Academies report from the Computer Science and Telecommunications Board will address the interface between computer science and biology.

termination and genome sequencing—there was no intent that the committee analyze specific DOE programs or restrict itself to DOE's existing programmatic boundaries. Hence, the recommendations are stated in general terms and are applicable to programs at any of the funding organizations whose missions encompass the mathematical sciences, biology, and the interactions between these fields, including but not limited to DOE. The committee has worked very hard to provide substantiated guidance about the scientific opportunities that these organizations are poised to support.

This report has been reviewed in draft form by individuals chosen for their diverse perspectives and technical expertise, in accordance with procedures approved by the NRC's Report Review Committee. The purpose of this independent review is to provide candid and critical comments that will assist the institution in making its published report as sound as possible and to ensure that the report meets institutional standards for objectivity, evidence, and responsiveness to the study charge. The review comments and draft manuscript remain confidential to protect the integrity of the deliberative process. We wish to thank the following individuals for their review of this report:

James Collins, Boston University,
Terry Gaasterland, Rockefeller University,
David Haussler, University of California at Santa Cruz,
Douglas Lauffenburger, Massachusetts Institute of Technology, and
Simon Levin, Princeton University.

Although the reviewers listed above have provided many constructive comments and suggestions, they were not asked to endorse the conclusions or recommendations, nor did they see the final draft of the report before its release. The review of this report was overseen by Ronald Douglas, Texas A&M University. Appointed by the National Research Council, he was responsible for making certain that an independent examination of this report was carried out in accordance with institutional procedures and that all review comments were carefully considered. Responsibility for the final content of this report rests entirely with the authoring committee and the institution.

In addition, the committee thanks Mark Daly, Avner Friedman, and Alan Perelson for their remarks and suggestions during the study process.

Contents

Executive Summary

The exponentially increasing amounts of biological data at all scales of biological organization, along with comparable advances in computing power, create the potential for scientists to construct quantitative, predictive models of biological systems. Broad success would transform basic biology, medicine, agriculture, and environmental science. The main push in biology during the coming decades will be toward an increasingly quantitative understanding of biological function; the rate at which progress occurs will depend on a deeper, effective implementation of quantitative methods and a quantitative perspective within the biological sciences.

The success of this transformation will depend in part on the creation and nurturance of a robust interface between biology and mathematics, which should become a top priority of science policy. The policy challenges will be substantial and multifaceted. The interface between biology and mathematics is an interdisciplinary frontier sprawling across a vast expanse of intellectual terrain that is extraordinarily diverse, indistinctly marked, and growing. The committee will explore this frontier in the chapters that follow. While it is not possible to capture all of the terrain in a single study, the committee attempted to identify striking features that exemplify the opportunities and also the challenges.

RECOMMENDATIONS

The committee offers five recommendations.

Recommendation: Funding agencies supporting mathematical research related to the life sciences should be receptive to research proposals that pertain to any level of biological organization: molecules, cells, organisms, populations, and ecosystems. While much current research can be productively confined to a particular level, there are also substantial challenges and rewards associated with analyzing interactions between levels.

The biological sciences are already becoming more quantitative and data-intensive; indeed, the explosion of data production and the potential for quantitative analysis replete with estimates of precision are the most visible qualities of the biological sciences of the 21st century. Progress in the biosciences will increasingly depend on deep and broad integration of mathematical analysis into studies at all levels of biological organization. No one level of organization stands out as offering singularly attractive opportunities for mathematical applications. The challenges faced at different levels have distinctive characteristics, but there are also unifying themes. Some chapters of the report are organized around the different levels of biological organization, but others—including "The Nature of the Field," "Historical Successes," and "Crosscutting Themes"—look more broadly at the commonalities of past and current applications of mathematics to biology.

Recommendation: Funding agencies supporting mathematical research related to the life sciences should give preference to proposals that indicate a clear understanding of the specific biological objectives of the research and include a realistic plan for how mathematicians and biologists will collaborate to achieve them.

The committee regards the interface between mathematics and biology as *biology-driven*. Research that proceeds by abstracting biological problems away from specific biological contexts and explores the properties of the resultant abstraction is less likely to be effective than research that stays more tightly focused on actual biological questions. However, to maximize productivity, the most powerful and appropriate mathematical tools should be selected to address important biological problems, and this quest benefits from involving the dedicated expertise of mathematical scientists. There are also many cases where results developed within pure mathematics, or in applications of mathematics to physical systems and engineering, later find powerful applications to biology, but this process, too, is most productive when it is biology-driven. Furthermore, the committee was impressed with the sheer scope of mathematical applications to biology and the diverse types of mathematics that are playing important roles in the life sciences. Hence, it strongly cautions against prejudging which subfields of mathematical research are most likely to contribute to biology.

> **Recommendation: Funding agencies supporting mathematical research related to the life sciences should give priority to research that addresses intrinsic characteristics of biological systems that reappear at many levels of biological organization: high dimensionality, heterogeneity, robustness, and the existence of multiple spatial and temporal scales.**

Biological systems at all scales are characterized by high dimensionality, heterogeneity, robustness to perturbations, and the existence of strongly interacting, highly disparate spatial and temporal scales. While these characteristics also appear in some physical systems that have been successfully modeled mathematically—the modeling of heterogeneous and multiscale phenomena is in particular a vibrant topic of mathematical and engineering research—the modeling of biological systems will require greatly expanded capabilities in these areas. As is widely documented in the report, the characteristics enumerated above recur at all levels of biological organization, from molecules to ecosystems.

> **Recommendation: Funding agencies supporting mathematical research related to the life sciences should support the refinement of general-purpose tools whose broad biological utility has already been established. Such research might require specialized review criteria, particularly when the focus is on tool enhancement rather than breakthrough research.**

Although the committee feels strongly that mathematical research based on premature abstraction of biological problems risks irrelevance, there are more and more instances where mathematical tools have already proven their utility in a broad range of biological applications. Many examples are described in this report. In some instances, as biological applications of these tools have expanded, limitations on their effectiveness have become apparent. Nonetheless, there are opportunities here for effective and important mathematical research that is less tightly tied to particular biological applications than is typically the case. Such research will have varying degrees of innate mathematical interest but can have an important impact on biology.

> **Recommendation: Funding agencies supporting mathematical research related to the life sciences should place increased emphasis on funding mechanisms and novel approaches to the**

organization of interdisciplinary research. The goal should be to foster effective collaboration between mathematical scientists and bioscientists by working to eliminate barriers posed by inadequate communication, disparate timescales for achieving research objectives, inequitable recognition of contributors to interdisciplinary projects, and cultural divisions within universities, research institutes, and national laboratories.

The committee's charge was to explore research areas at the interface between mathematics and biology that are likely to offer particular promise in the years ahead. Hence, it did not undertake a broad examination of funding mechanisms, training, and the organization of interdisciplinary research projects. However, these issues came up so frequently in its deliberations and are so central to the future prosperity of research at this interface that the committee recommends they receive increased attention. Given the many cultural factors that impede optimum collaboration between mathematical scientists and bioscientists, it would be desirable to explore a variety of mechanisms for overcoming them or minimizing their deleterious effects.

RATIONALE FOR THE RECOMMENDATIONS

The committee's recommendations are notable both for what they say and what they omit. There is, for example, no call in this report for a major initiative to develop an in silico cell or any other major, potentially multiagency initiative with a specified goal. A recommendation suggesting such a high level of administrative organization so singularly directed seems premature at the least and, the committee believes, would likely be counterproductive at this time. The committee opted for a patient, broadly based, vigorous effort to expand research at the interface between mathematics and biology rather than for a commitment to a small number of high-profile projects with monolithic goals. Because this decision was perhaps the most consequential outcome of the committee's deliberation, it is appropriate to summarize the committee's rationale.

The committee undertook this study at a time of dramatic change throughout the biological sciences. During the past decade, a "perfect storm" of developments has touched off broad changes in biology. Unlike most past discontinuities in the biosciences, this one was not triggered by major scientific discoveries: It was triggered instead by a confluence of new technologies that has swept broadly across science and society, as well as by developments internal to biology. Key contributors to this perfect storm include the following:

- *The development and widespread adoption of automated instruments that produce high fluxes of digital data relevant to all levels of biological organization.* These instruments have transformed DNA sequencing; analysis of mRNA and protein populations; determination of protein structures; structural and functional imaging of subcellular organelles, cells, tissues, organs, and whole organisms; electrophysiology; analysis of genetic variation in populations; and ecological changes across the entire biosphere.

- *The arrival of networked, high-performance computing systems on the desktops of all biologists.* This sudden access to computing resources—a phenomenon whose roots lie in the same technological revolution that enabled the development of the high-throughput instruments discussed above—has infused quantitative methods into all facets of biological research. High-performance computing impacts the whole range of research activities, from the low-level processing of raw data and the development of new theoretical frameworks to the organization, dissemination, and analysis of large biological databases.

- *The success of the Human Genome Project in establishing accurate, whole-genome sequences as central resources in biology.* Genome sequences have given biologists their first taste of "complete knowledge" and stimulated intensive efforts to improve our ability to recognize genes in genomic sequence, to discern their functions, and to infer their evolutionary histories. Genome sequences have also led to a renewed appreciation of the molecular unity of life. The conservation of nucleotide and amino acid sequences and the associated conservation of molecular functions have made sequence comparison the centerpiece of genome analysis. Hence, the analytical challenges in genomics are expanding with $O(n^2)$ complexity, where n is the number of known nucleotides in the DNA sequence—a number that is itself growing exponentially.

- *The maturation of a phase of molecular and cellular biology during which biologists acquired robust, albeit largely qualitative, descriptions of the basic molecular pathways that allow the self-replication and development of organisms and that govern their utilization of energy and interactions with their environments.* This great flowering of the biological sciences gained momentum following the discovery of the double helical structure of DNA. While much productive research continues to expand upon and refine the basic paradigms of the late 20th century, biology is also visibly in transition. Bioscientists in many research areas recognize the need for a more quantitative, integrated, and predictive understanding of living systems rather than a simple expansion of current modes of biological analysis to encompass ever more phenomena.

Collectively, these developments are transforming biology into a more quantitative, data-intensive science, a transformation that has important implications for the interface between mathematics and biology. As noted in the Preface, the committee interpreted "mathematics" broadly so as to include computational science and statistics, as well as all aspects of applied mathematics. Its study also spanned all levels of biological organization, from molecules to ecosystems. Given this breadth of view, it is not surprising that the committee is impressed by the richness and diversity of current research at the interface between mathematics and biology. That richness and diversity pose a substantial challenge to science policy. Certainly a strong case can be made for increased policy attention to this interdisciplinary frontier. However, in the committee's view, it would be unwise to channel this attention too narrowly into the pursuit of particular high-profile opportunities.

The tension between diversified efforts to strengthen important research areas and the commitment of major resources to high-profile projects is an enduring feature of the science-policy landscape. The Human Genome Project was perhaps an ideal model of a successful high-profile project. Much as envisioned by the National Research Council report *Mapping and Sequencing the Human Genome,*[1] it led to a flourishing of technical advances in our ability to analyze DNA and a much closer connection between research on model experimental organisms and human biology. Too, it culminated in the generation of reference databases that have become indispensable tools for everyday research on many organisms. These databases have also become critical frameworks around which expanding knowledge of molecular and cellular processes can be rationally organized. The Human Genome Project introduced high-throughput technology to the biological sciences, which in turn led to a profound change in how biological research is conducted and also to the data-rich world where biologists now work that enables the introduction of more quantitative approaches. On a philosophical level, the Human Genome Project was big science in the service of small science. Throughout its history, the project empowered rather than displaced small laboratories as the engine of biological innovation, and in its aftermath it continues to do so.

Other science-policy initiatives that made more equivocal contributions also provide historical context for the committee's recommendations. For example, the War on Cancer of the late 1960s is often cited as a misguided effort to concentrate resources on an ill-defined goal toward whose achievement contemporary science offered no clear path. The mainstream

---

[1]National Research Council, *Mapping and Sequencing the Human Genome*, National Academy Press, Washington, D.C., 1988.

verdict on the War on Cancer holds that patient, diversified support for molecular and cell biology would have been more appropriate than a high-profile project organized around the easily articulated, practical theme of eradicating cancer. However, even an initiative such as the War on Cancer, much of whose rhetoric appears embarrassing in retrospect, can be a powerful stimulant of needed changes in scientific priorities. There is little doubt that the War on Cancer accelerated the diversification of molecular biology beyond its bacterial roots, helped lay the foundation for the recombinant-DNA revolution, and brought basic biology into closer partnership with medicine. There were also collateral benefits from the War on Cancer's vigorous pursuit of a largely incorrect hypothesis, which was that retroviruses were a major cause of human cancer. The War on Cancer encouraged development of experimental techniques for isolating and growing retroviruses and expanded knowledge about their life cycles, which proved invaluable in confronting the AIDS epidemic. Nonetheless, the committee does not believe that the possibility of collateral or unexpected, unplanned, perhaps serendipitous contributions from a high-profile project would be an effective way to bring quantitative methods into the biological sciences and quantitative descriptions into our understanding of biology.

In considering research opportunities at the interface between biology and mathematics, these historical precedents—the Human Genome Project and the War on Cancer—influenced the committee's thinking about a key policy question. Should funding agencies channel resources into grand challenges such as these as a way of stimulating interactions between mathematics and biology in a big way? It is not difficult to identify candidates for such grand challenges at all levels of biological organization: They would include the development of a comprehensive, predictive computer model of a particular free-living cell, organ, or ecosystem. At various levels of ambition, such initiatives are already under way. It is also not difficult to see that the rapid expansion in biological data requires a multiplicative, rather than merely an incremental, expansion in the number of researchers working on mathematical aspects of biology. In fact, a narrowly defined, high-profile project like the two featured above might slow the overall introduction of quantitative methods into the biological sciences, might retard the general training of biologists in more quantitative methods, and might not develop the range of mathematical applications that could transform many areas of biology. Mathematical scientists and methods tuned to that particular grand challenge would, of course, be greatly encouraged and benefit directly, and bioscientists involved in the project could come to appreciate the role of mathematics. However, the science-policy dilemma is whether or not biology is best served at this time by the type of organized multiagency, multi-investigator coordination—and focused infusion of resources—that made the Human Genome Project a success. If defined ambitiously, an all-out effort to create a predictive computer model of a free-living cell—or any similar project at other levels of biological organization—would require an enormous concentration of experimental, theoretical, and computational resources around a well-specified, centrally sanctioned goal.

While recognizing the potential of such an initiative to stimulate coordinated action in an underdeveloped research area, the committee opted instead to recommend a long-term, broad, and diversified nurturance of the interface between mathematics and biology. Two themes that recurred during the committee's deliberations influenced this choice—the primacy of the biology problem and the lack of predictability.

Primacy of Biology

In applications of mathematics to biology, the committee returned again and again to the primacy of the biological problem. The primary goal of funding agencies and researchers working at the interface between mathematics and biology should be to solve particular biological problems, not to accomplish particular feats in the mathematical description of living systems. Hence, an all-out effort to "understand" the bacterium *Escherichia coli* or the yeast *Saccharomyces cerevisiae*, if undertaken, should have biological goals. Perhaps a predictive computer model is part of what is needed, but it should not be the central goal. Indeed, computer modelers participating in such projects should be guided by the biological objectives. Some modeling approaches will be more appropriate to particular objectives than others. Both biological progress and mathematical progress are likely to be optimized by intimate coupling of whatever modeling is done to defined biological objectives. Implicit in this view is the committee's sense that we are far away from having an in silico cell. A very large amount of experimental bioscience research would be a prerequisite for the modeling, and a wide range of subcellular elements with their own daunting complexities might well have to be tackled first, both to provide models or prototypes and test beds and to facilitate understanding what is needed and what can be ignored in constructing a successful in silico model of a cell. An analogy with the history of artificial intelligence research may help clarify the committee's thinking. One could envision a "Turing test" for the in silico cell. To conduct such a test, an experimentalist would design manipulations and measurements to be carried out on the target cells; results would then be returned based on the experimental manipulation of real cells on the one hand and their computer simulation on the other. For the simulator to pass the Turing test, it should be impossible for the experimentalist to devise manipulations and measurements that would distinguish between the two sources of data. Grand as this challenge might be for 21st century biology, we are too far from meeting it for it to be a dominant organizing principle for current research. While some efforts in this direction are clearly worthwhile, it should be kept in mind that premature efforts by the artificial intelligence community to pass the original Turing test floundered. So, too, would a present-day effort to meet the corresponding challenge of an in silico cell. Progress in artificial intelligence has depended on breaking the ultimate task into many smaller, more accessible tasks, each of which is approached with a variety of strategies. Similarly, the committee concluded that contemporary biology would be best off adopting an incremental and diversified approach to the creation of more quantitative, predictive descriptions of living systems.

Unpredictability

The committee found the history of applications of mathematics to biology to be full of unexpected turns and reciprocal influences on the two fields and expects this dynamic to continue. Success in big-science initiatives depends on an element of predictability about how areas of science will develop. Certainly there was technical risk that the Human Genome Project would prove premature, and there was even some risk that genome sequences would prove so difficult to interpret that their impact on biology would be minimal. Nonetheless, by the late 1980s, it was abundantly clear that DNA sequencing was capable of providing much useful information about biology, that there were open-ended opportunities to lower its cost and increase its throughput, and that genome sequences would play a very important role in the future of biological research. Similarly, it was apparent by the late 1960s that the successes of the first decades of molecular biological research on bacteria should be extended to eukaryotic and metazoan organisms. The committee is less confident that the future directions of the interplay between mathematics and biology can be reliably predicted in 2005. While it is confident that mathematical methods will become steadily more deeply integrated into biological research, the committee regards the directions in which the biological sciences will evolve in the decades ahead—and the detailed ways in which mathematics will facilitate that evolution—to be highly uncertain. The excitement that surrounds this area of scientific research stems from a blend of opportunity and unpredictability. Many areas of biological research are at points of instability. The ways in which these instabilities resolve will shape the future of the relationship between mathematics and biology.

FUTURE PERSPECTIVE

The committee is confident that deepening interactions between mathematics and biology will transform the biosciences. Of equal interest is the possibility that the areas of mathematics that interact most strongly with biology will themselves be altered by these interactions. Indeed, much of modern mathematics was shaped by four centuries of intimate interaction with the physical sciences and engineering. As the prominence of the biosciences increases—and as they interact more intensively with mathematics—a similar dynamic may be expected to occur. As discussed above, biological processes have different characteristics than the processes commonly encountered in engineering and physical science. In comparison with scientists involved in materials science, plasma physics, or cosmology, bioscientists work on muddier problems. The vast scales of time and space that characterize the world of biology are complemented by nonquantitative, organizational features that are so extraordinarily complex. The number of different interacting components is huge (ranging up to millions or even billions of entities), and they can all possess individual characteristics and contingent properties and be influenced by historical events. The systems are typically far from equilibrium or even stable steady states. High-order interactions between the components are the rule: The amount of feedback regulation in the simplest cell greatly exceeds that presently incorporated into devices designed by humans. (Indeed, it is this reliance on feedback regulation that accounts for the robustness of living systems.) Small events at one spatial or temporal scale often have large effects at another very different scale. These generalizations apply to cells, whose components are molecules, and also to ecosystems, whose components are commonly taken to be populations of individual members of many species. Calculus, the mathematical properties of continuous, very small elements, has been the essential language for describing the physical world and the language employed in the physical sciences, but biology has discrete elements, and the quantitative language of the computational and information sciences appears far more suited to be the language of biology. As a consequence of these many ways in which biology differs from the physical sciences, the committee looks forward to its many influences on mathematics, including some explicitly new mathematics.

An important goal in developing this report was to illustrate, in diverse contexts, these distinctive characteristics of biological systems. They may appear intimidating to nonbiologists at first, but on closer inspection, it is apparent that there has been great progress in dealing with them in the past and that this process is expanding as more mathematicians address biologically motivated problems. Historically, some new mathematics simply emerged from the inner workings of the human brain, without the direct influence of external reality. However, many of the finest moments of pure and applied mathematics have arisen in response to humankind's quest to understand the physical world. The committee believes that the 21st century's intensifying quest to understand the living world will provide an equally rich stimulus for future triumphs.

1

The Nature of the Field

INTRODUCTION

Biology is in dramatic flux due to a surge of new sources of data, access to high-performance computing, increasing reliance on quantitative research methods, and an internally driven need to produce more quantitative and predictive models of biological processes. The growing infusion of mathematical tools and reasoning into biology may therefore be expected to further transform the life sciences during the decades ahead. This transformation will have profound effects on all areas of basic and applied biology.

Nonetheless, we are not starting from scratch in applying the mathematical sciences to biology. To a greater extent than is widely recognized—by biologists and nonbiologists alike—there has been a string of dramatic successes over more than a century that have been critical to advances in biology and have also led to new mathematics. The role that biological problems played in motivating the development of modern statistics is just one example that will be described in Chapter 2, "Historical Successes."

THE MATHEMATICS-BIOLOGY INTERFACE

The interface between mathematics and biology can be examined across scales of biological problems and across all the major areas of mathematical sciences. Biological scales range from molecules, cells, organisms, and populations to communities.[1] Much of the remainder of this report is organized around these biological scales, articulating examples of the biological problems to be addressed at each scale. These scales may be briefly described as follows:

- *Molecules.* Molecular biology focuses on the chemical components of life and their interactions. These components differ greatly in size and complexity, ranging from atoms and simple ions, through the basic molecular building blocks of life such as nucleic and amino acids, sugars, and fats, to polymers and homogeneous and heterogeneous aggregates of the more basic units, forming macromolecular assemblies and supermolecular structures that carry out many of the fundamental processes in the life of a cell. The structures of these objects, as well as the dynamics of molecules and interactions between them, are central to biological function.

- *Cells.* Cell biology is concerned with the self-replicating units of life, including bacteria, plant, and animal cells, as well as the viruses and other parasites that infect them. The study of the cell also includes consideration of many interconnected units or subcellular structures, such as organelles, which range in complexity from peroxisomes, proteosomes, or lysosomes, to mitochondria and chloroplasts, up to the nucleolus and the nucleus itself for eukaryotic organisms, and other structural components intrinsic to cell function such as the endoplasmic reticulum. The mechanisms and consequences of cell–cell communication are also of primary interest.

- *Organisms.* Organismal biology includes both the properties of whole organisms and the complex multicellular structures of which they are composed—the tissues, organs, organ systems, and integrative processes that create a robust whole out of diverse parts. Organisms sustain health and well-being in the face of considerable insults and environmental disturbances, a process known as homeostasis. Another feature at this scale is the study of the breakdown of this robustness—in other words, the etiology and nature of disease.

---

[1]Dividing life into levels, or scales, is obvious, is essential for understanding, and reflects an intrinsic feature of biology. Nonetheless, the levels interact, and some of the division is for human convenience or is an artifact of scholarly history. Characterizing any one level requires at least considering its immediately adjacent levels; one could also provide a finer subdivision of some of the scales, but for clarity, the committee used the most commonly employed and obvious distinctions, ones that are important for how biologists think about the object of their study and that provide a means for mathematicians to think about how to engage biology.

- *Populations.* Population biology concerns groups of organisms of the same species. Genetic variation among individuals is of primary interest, as is the behavior of populations over time in real environments—for example, speciation, population fluctuation, and extinction.

- *Communities and ecosystems.* Community ecology is the study of assemblages of populations of different species and their interactions. Interactions between living and nonliving components, nutrient and carbon fluxes, and overall responses of ecosystems to changes in the physical environment are central issues of ecosystem ecology. The ecology of infectious disease is an example of an interspecies interaction of considerable recent interest.

Across all these levels of biological organizations, modeling of biological processes plays a central role. Much of this report describes the types of models—and the associated mathematical techniques—that have been productive in biology. In considering this diverse landscape, it is important to understand that the word "model" has many meanings in biology. Concepts in biology are often illustrated by simple verbal or visual models that are entirely qualitative. For example, most models of gene-regulatory circuits are of this nature: They specify, in simple drawings, which components of a pathway inhibit, and which stimulate, either their own synthesis or that of other components. Other models may be formulated mathematically even though they are primarily intended for heuristic use rather than for data analysis: Simple differential equations describing idealized predator-prey interactions are in this category. Finally, sophisticated models designed to capture subtle features of large, real data sets are also diverse. Some are sets of partial differential equations that would look familiar to a classical physicist. Others are designed to capture subtle statistical properties of data sets without reference to operative biological mechanisms. Many are stochastic models that guide sampling from combinatorially explosive sets of possible relationships between biological objects: Examples include coalescent models of possible phylogenetic relationships between DNA sequences or between organisms defined by sets of discrete phenotypes. Hidden Markov models of sites of transcription-factor binding in the regulatory regions of genes are also in this category. Throughout the report, this diversity of models should be kept firmly in mind. Of course, diversity in types of models is found in all fields of science. Nonetheless, biology is perhaps unique in the extent to which diversity in modeling practices is the rule, with the existence of a small set of standard paradigms that are applicable to broad sets of problems being the exception.

WHAT HAS CHANGED IN RECENT YEARS?

While scientists have been studying biological systems at these various scales for many years and applying varying levels and types of mathematics and statistics to them, recent achievements in biology and technology have combined to create a dramatically new world of opportunity for the application of mathematics to biology. Rather suddenly, new experimental methods and technologies have allowed the generation of biological information at an astonishing rate. This phenomenon is playing out at all scales of biological analysis:

• On the molecular scale, the human-genome sequence and the sequences of many other genomes have been determined; methods are also available to measure the expression levels of all genes in an organism in a single experiment. New techniques in protein chemistry, as well as new radiation sources for structural analyses, are accelerating the rate at which proteins can be detected in complex biological samples, purified, and characterized structurally at atomic resolution.

• On the cellular scale, new methods of cellular imaging are making it possible to track subcellular processes and to trace the propagation of signals at millisecond timescales.

• For organisms, reductions in the cost of noninvasive imaging such as computed tomography (CT) scanning, magnetic resonance imaging (MRI), and positron emission tomography (PET) are making these methods available as routine experimental tools. New methods, such as high-throughput patch-clamp studies, are providing electrophysiology data at previously unattainable rates.

• For populations, it is now possible to measure the genetic differences between organisms at hundreds of thousands of sites in a single, inexpensive experiment.

• At the level of communities, new remote-sensing technologies are making it possible to measure entire ecosystems across multiple observation channels at high resolution.

The data that guide biology are diverse, and their integration is challenging. Data sets span the entire range from genomic data to satellite data. Data may be collected at one point in time or continuously, resulting in a real-time data stream (Turner et al., 2004; Running et al., 2004). In addition, there are some cases (e.g., the National Science Foundation (NSF)-funded Long-term Ecological Research Sites or the Framingham Heart Study) where data have been collected about biological entities over long periods of time.

Biological data may contain large errors of unknown origin owing to complex interactions that are either poorly understood or inherently due to stochastic processes. Quantities may be inferred using proxy data—for instance, seasonal or annual temperature in the past can be inferred from tree rings. Model development and model parameterization must take these sources of uncertainty—complex interactions and stochastic processes—into account.

In the past, data were almost exclusively collected by individual researchers or small groups of researchers, and they remained the property of those who collected them. Now, data are increasingly collected by larger groups of scientists and made publicly available. The Protein Data Bank and the National Center for Ecological Analysis and Synthesis are two examples of institutions established to promote the sharing of biological data. One impediment to synthetic analytical work that relies on large data sets collected by different groups has been the lack of commonly accepted standards for collecting, archiving, and annotating data and of agreement on what kinds of data should be collected. Nonetheless, there are increasing efforts to come to some agreement. For instance, Ecological Metadata Language (EML) is a way to standardize data annotation that is increasingly embraced by the ecological community.

Some of the greatest computational challenges today come from data collected at the two extreme ends of spatial scales: genomic data and satellite data. The explosive rate of genomic data generation is well known. In the case of more global data, Palumbi and colleagues (2003), for example, present a number of new data acquisition methods, including remote sensing to measure characteristics of the ocean (temperature, wind, surface elevation) or trace changes in ocean currents and DNA sequencing to assess spatial and temporal trends in genetic diversity. New computational methods need to be developed to extract useful information from these vast amounts of data.

The analysis of spatial data poses particular challenges due to the correlations that are inherent in spatial processes and due to local interactions and stochastic effects. As an example, a method widely used for detecting anomalies in space is the spatial-scan statistic. This statistic has been applied to the detection of disease outbreaks or invasive species (Patil and Taillie, 2003). One challenge is to couple on-the-ground observations and remotely sensed data. Visualization tools are indispensable when analyzing spatial data.

In parallel with the accelerating rate of data acquisition, there has been an increase in the computational power available to the scientific community—on the desktop, in a research unit, or through the Internet, from a national resource or a grid of independent systems. Before computers were widely available, discoveries were made by combining data from experiments or observational studies and their statistical analysis with simple models that served as the conceptual framework. When computing power became available, this paradigm was extended to include computation. This process started in areas of biology that are closely aligned with the physical sciences (e.g., protein structure determination) and gradually spread throughout biology. Concomitantly, biologists became increasingly dependent on sophisticated data-analysis tools and complex, data-driven models.

When analytical models are improved to the point of being good representations of a biological system, they are often analytically intractable, and biologists must turn to computation. In many cases, such models are systems of differential equations, which are fairly amenable to solution on computers thanks to mathematical advances of the past few decades. The numerical analysis community continues to increase the set of partial differential equations for which reliable, fast solvers exist. Robust techniques are now available for solving problems in electrostatics, diffusion, elasticity, and fluid dynamics. However, the needs of biology are among the most challenging. Particularly because it often calls for multiscale models, which include both deterministic and stochastic elements, the solution of sets of biologically motivated equations frequently exceeds current capabilities. There are not yet adequate capabilities for evaluating the range of uncertainties embedded in a computational model due to its parameters, discretization, and structure. Few tools are available to deal with these issues when the models are applied to large systems. Mathematical methods are also being used in new ways to inform experimental design. Traditionally, experimental design decisions, such as choosing the nature of a perturbation, the response measurements, whether to or how to do gene disruption, and the timing and scope of response measurement, have been made by the experimental biologist with only minimal consideration of the computational analysis that will be performed based on the resulting data. This was perhaps unavoidable, as techniques for gene disruption and high-throughput assays have until recently been the major limiting factors. However, as experimental genomic science advances, options are becoming increasingly available. In the future, experimental design considerations must be tightly coupled to the mathematical representations to be used to model the system and the computational and statistical methods to be used for model identification and parameter estimation.

For example, switches for transcription or protein modification have recently become available (Shimizu-Sato, 2002; Zeidler et al., 2004), making it feasible to implement oscillatory perturbations in a systematic manner. A natural question arises: Does oscillatory perturbation have any advantage over traditional impulse or step perturbation? And if so, how do we quantify the advantages? A related question is what mathematical and statistical techniques are useful for the analysis of the response to such oscillatory signals. To answer these questions, we need to study how to estimate the system parameters based on each type of perturbation and how the estimation may be affected by intrinsic and measurement noise. The mathematical techniques involved include Laplace transforms, optimal time and frequency sampling, and estimation theory under ill-posed conditions. Even for a stochastic network as simple as an autoregulatory gene whose protein activities are modulated by an input signal, these questions have not been studied until very recently (Lipan and Wong, 2005), and much more remains to be done. Although the benefits of oscillatory perturbations are widely appreciated for the analysis of physical systems, this approach is rarely used in the study of cellular systems. Careful mathematical and simulation studies may help to interest experimental investigators in evaluating the promise of novel perturbations of biological systems.

From the point of view of software, there are a number of open issues that are common to many scientific disciplines. These issues include the fact that codes are complex and difficult to support, especially on parallel computers (Post and Kendall, 2004). Software architectures that allow problem specification and access to all layers of code development would constitute an important step forward. Message passing interface (MPI)-based development platforms such as Portable, Extensible, Toolkit for Scientific Computation (PETSc) can accelerate the rate of progress, and efforts like DOE's Scientific Discovery through Advanced Computing (SciDAC) program are promising. An upcoming report from the National Academies' Computer Science and Telecommunications Board will discuss, in particular, the interface between biology and the computing world.

Another issue is verification and validation. Verification is defined as determining whether the calculations truly correspond to the equations that constitute the analytical model. There are well-defined techniques for verification, but they are rarely used systematically. Validation is a broader concept that is generally understood to mean the assessment of the model quality—that is, does the software correspond to biological reality? A part of the validation process that is common in the physical sciences but little used in biology is the conduct of experiments designed specifically to test the computational model itself rather than to study new phenomena. That is, we need to consider the output of a computational model as a testable hypothesis and then design biological experiments that try to disprove the hypothesis by collecting appropriate data or exploring whether qualitative features of the computer output exist in real systems. In order for such approaches to contribute significantly to progress in understanding biology, the experimental community will need to be convinced of the value of such studies, which may not directly address the experimentalists' goals.

WHAT MAKES COMPUTATIONAL BIOLOGY PROBLEMS HARD?

While the challenges posed by rapidly increasing amounts of data cut across all the sciences, those challenges posed by increased amounts of data in biology are uniquely difficult. At all scales of analysis, biology involves large numbers of types of objects, large numbers of objects of each type, and complex interactions between objects. In addition, biological objects can possess individuality, a history (e.g., of external stimuli, environmental insults, or inheritance), and a contingent existence (e.g., the location of components or of neighbors can be significant.) Except for relatively small contributions from phenomena such as bilateral body plan (when and where relevant), schemes for simplification that arise from symmetry are rarely possible at any scale in biology. The systems are extraordinarily heterogeneous in space and time, yet stunningly robust in the face of perturbation. Interactions across vastly different scales can have dramatic effects on system behavior. Thus, a tremendous quantity of data must be managed in creating useful biological models. Moreover, some of those data are very difficult to obtain. These issues are described in more detail below for the molecular scale; similar issues arise at the other scales and across scales:

- *Large number of types of objects and objects of each type.* At the molecular scale, with minor exceptions, proteins are synthesized from only 20 different amino acids. These units combine to produce tens of thousands of independently encoded proteins in humans, and there are many different mechanisms that can lead to the creation of variants—sometimes astonishing numbers of variants—of each of these independently encoded proteins. Analogous phenomena occur with nucleic acids and the polymers of sugars, fats, and other molecules.
- *Complex interactions between objects.* DNA and RNA interact intramolecularly and with each other. DNA is the template for creating RNA, and RNA is the template for proteins. RNA and protein combine to form superstructures that themselves play central roles in the translation of RNA into proteins. Proteins interact intramolecularly and with other proteins, as well as with RNA, DNA, and a large variety of other molecules to act as enzymes, structural components, signals, receptors of signals, and inhibitors of signals.
- *Robustness.* The many types of molecules in biological systems combine to form extraordinarily robust subcellular organelles, cells, tissues, organs, organ systems, organisms, populations, and communities. Bio-

logical interaction networks achieve this robustness through high levels of redundancy, modularity, heterogeneity, and feedback. It is not uncommon to find that genetic ablation of a normally critical signaling pathway not only fails to kill a cell but causes only the most subtle changes in its behavior. Other pathways can often provide similar functions even if they do not normally do so when the primary pathway is present. Similarly, feedback mechanisms make biological systems extraordinarily robust against both internal and external perturbations. Genetic variation is a particularly important example of an internal perturbation. Between the two copies of the genome present in each human, there are millions of sequence differences, many of which affect the regulation of genes and the structures of the encoded proteins.

• *Complex interactions across scales.* All of this complexity is present at each scale of organization and in the interactions between scales. For example, it is infeasible to simulate organ-scale electrophysiology by modeling the ion fluxes through every membrane channel in every cell. There is, at the present time, no systematic way of bypassing this problem. Existing approaches tend to be hybrid methods, which overcome such bottlenecks by using different models on different temporal or spatial scales coupled with heuristic models to transfer information between them. Statistical methods are also used to integrate information obtained from fine-scale calculations to estimate the net response of an organ, tissue, or neural network.

FACTORS COMMON TO SUCCESSFUL INTERACTIONS BETWEEN THE MATHEMATICAL SCIENCES AND THE BIOSCIENCES

As the committee examined the historical record and contemporary experience in applying mathematics to biology, a few simple observations that commonly underlie successful interactions came to the fore:

• The biological problem has always been primary. Successful applications of mathematics to biology are driven by a deep understanding of the relevant biology. Until this understanding is in place, it is not possible to state the problem with sufficient clarity and at a sufficient level of abstraction to allow a meaningful mathematical formulation. Successful applications always involve major simplifications of the actual system. However, these simplifications must preserve the system's essential features. This first observation gives rise to the following recommendation:

Recommendation: Funding agencies supporting mathematical research related to the life sciences should give preference to proposals that indicate a clear understanding of the specific bio-

logical objectives of the research and include a realistic plan for how mathematicians and biologists will collaborate to achieve them.

There are dual benefits of preferential support for proposals rich in both biological understanding and clarity about the mechanisms by which the collaboration will advance. Naturally, well-organized and well-posed research aimed at important biological problems will pay off early on, will help sustain further studies, and will open up new directions for fruitful inquiry. In addition, establishing such preferential support minimizes the risk of applying mathematics to poorly posed biological problems and maximizes the potential impact of quantitative tools. Rigorous prioritization will support a structural change in the biological sciences that encourages the use of quantitative approaches of all categories. More generally, success stories based on such considerations will be readily exported to other biological research problems and will serve to validate the role of mathematicians in biology qua mathematicians rather than just as technical contributors and to validate, for experimental biologists, the role of mathematics itself in understanding biology.[2]

• As the committee discusses in more detail below, cultural and linguistic barriers create a potentially large divide between mathematicians and biologists. It is only after achieving a common language in which to discuss a particular problem that mathematics can be applied effectively. The common ground can lie anywhere along the spectrum from the language of biology to that of mathematics, but it has to be found, and each side has to move toward the other to do so. That said, it is important to recognize that communication barriers that appear to be linguistic often have deeper roots. Many of the difficulties that researchers trained in the physical sciences, engineering, and mathematics have in communicating with biologists relate to fundamental differences between biology and the physical sciences. Basic laws typically do not exist, and even basic principles are often still undiscovered. Once they understand that progress is possible despite these obstacles, some nonbiologists thrive in this strange, new scientific environment. Others find that their skills are best applied in better-defined settings.

• Initial progress has almost always depended on existing mathematical tools, often quite elementary ones. The complexity, particularly at early stages of analysis, is in the biology, not the mathematics. Any improvements to mathematical tools come later.

---

[2] Of course this gap does not exist if one individual is well grounded in both fields. However, it is more common, and generally more practical, to collaborate rather than to learn two disparate fields.

• Formulation of the problem has been as important as solving it. As they are first formulated, biological problems are typically ill posed or incompletely posed. The process for translating them into formal statements in the language of mathematics introduces a rigor that often uncovers questions that might not otherwise have been asked. The translation process causes both bioscientists and mathematicians to think carefully about all of the parts of the system and to decide systematically which variables, effects, and interactions to take into account. This process is also a critical test of whether the biologists and mathematicians working together on a problem have actually arrived at a common language.

• Even though many biological problems have been solved using simple mathematics, a sophisticated and experienced mathematical scientist has often been required to find the solution. This paradox arises because of the difficulty of abstracting the problem from its biological messiness and sifting through the enormous collection of tools and methods already potentially available for addressing it. In addition, the solution often involves applying familiar mathematical methods in unfamiliar ways or contexts.

There have been cases where mathematical techniques were applied to biological problems with inadequate appreciation for the finer points of the biology, leading some to overstate the significance of their mathematical results. The result was statements such as that of Mayr (1982, p. 304), who when explaining the role mathematics played in evolving the thinking of the ancient Greeks, wrote "This was the first of countless episodes in the history of biology where mathematics or the physical sciences exerted a harmful influence on the development of biology." This notion has held back the full introduction and exploitation of the power of mathematics in the study of biology. At the same time, a healthy skepticism is necessary for making progress in the sciences, and too-universal acceptance of approaches can impede progress as much as outright rejection. A balance of different approaches often yields the greatest gains, as eloquently expressed by Naeem (2002) in the context of ecology: ". . . ecological truth lies in the confluence of observation, theory, and experiment. It is through discourse among empiricists and theorists that findings and theory are sorted and matched and where there is a lack of correspondence, new challenges identified."

PREPARING THE GROUND FOR IMPROVED SYNERGIES OF BENEFIT TO BOTH FIELDS

Progress in the life sciences will increasingly depend on deep and broad integration of mathematical analysis into the study of all levels of biological organization. No one level of organization stands out as offering singularly attractive opportunities for mathematical applications. The challenges faced at different levels have distinctive characteristics, but there are also unifying themes.

> **Recommendation: Funding agencies supporting mathematical research related to the life sciences should be receptive to research proposals that pertain to any level of biological organization: molecules, cells, organisms, populations, and ecosystems. While much current research can be productively confined to a particular level, there are also substantial challenges and rewards associated with analyzing interactions between levels.**

The empirical factors for success listed in the previous section all point to one critical element: A true collaboration that brings together skills from the mathematical sciences and a deep knowledge of biology must be established. In response to this basic need, funding programs, research institutions, and groups can experiment with conditions to facilitate such an establishment. Some of the factors to be addressed include these:

- *Communication.* It is clear from the above that mathematical scientists and biologists have to find a common language so that all of the essential richness of a biological problem can be captured and formulated in mathematical terms. This can and should happen in both directions, with some biologists developing a deeper and more sophisticated understanding of quantitative methods and many mathematical scientists expanding their understanding of biology to appreciate the scope of the problems to be addressed. (The primary model in the mind of the committee is mathematical scientists contributing to biology research teams, not, for the most part, biologists learning all the necessary mathematics and statistics.) Interestingly, some of the most successful practitioners at the interface have come out of the physical and mathematical sciences, bringing a deep understanding of quantitative methods as well as biology, but neither to the exclusion of the other.

- *Timescales.* The professional timescales of the fields are often mismatched, and both sides of the collaboration need to develop an appreciation for this reality. On the one hand, if a biological challenge demands the development of deep new mathematics or statistics, this process will typically require a detour of months or years, time that is not consistent with the competitive nature of researchers in the biological sciences and the expectations of them. On the other hand, existing mathematical methods might require the generation of additional data (e.g., to enable good bounds on parameters or uncertainties), which might be time consuming and initially unrewarding to biologists.

• *Recognition and advancement.* If mathematical scientists are to invest time and effort in learning biology and to contribute what, from a mathematical perspective, may be relatively simple methods, then the mathematical sciences must adjust their reward systems. This difficulty is an age-old problem in academic departments: It flares up as practitioners in a field venture out to the interface with another field and devote more intellectual energy to transitioning research results than to directly advancing their own field's research agenda. Of course, university departments will not adjust something as fundamental as their own internal reward system in the absence of external stimuli and external rewards. While simply putting forward funds for collaborations at the interface will provide some incentive, the funding agencies need to consider special honorific awards and special programs, and possibly other mechanisms, to encourage the needed changes in systems for recognition and advancement. Adjustments would also help with the differences in timescales between the expectations and realities of doing biology and doing mathematics, and agencies could consider mechanisms to satisfy both timescales. Provision of more funding at the interface, as planned, is the first step.

> **Recommendation: Funding agencies supporting mathematical research related to the life sciences should place increased emphasis on funding mechanisms and novel approaches to the organization of interdisciplinary research. The goal should be to foster effective collaboration between mathematical scientists and bioscientists by working to eliminate barriers posed by inadequate communication, disparate timescales for achieving research objectives, inequitable recognition of contributors to interdisciplinary projects, and cultural divisions within universities, research institutes, and national laboratories.**

In spite of the committee's belief that most problems in biology can initially be addressed with fairly standard mathematics or statistics, there are occasions where exceptionally innovative researchers may be driven by the particularities of a problem to break out of traditional mathematics paradigms and develop truly novel methods. R.A. Fisher's work on the analysis of variance is a dramatic example addressed in Chapter 2, "Historical Successes."

There are also many examples where interesting mathematical problems were abstracted away from the biological problems that motivated them, leading to mathematical sciences research that is valuable in its own right. Examples of this type are particularly common in combinatorics, algorithmics, and computational complexity theory. A typical example is the "adjacent ones" problem, which first arose in the 1950s in the context of fine-structure genetic mapping. Once posed, it continued to interest mathematicians—and occasionally found new biological applications, including in the Human Genome Project—for 40 years (Benzer, 1959; Alizadeh et al., 1995). The committee's sense is that the flow of research problems from biology back into mathematics is likely to become increasingly common as research expands at the interface of the two fields.

It is important that more biologists recognize the value of true collaboration with mathematical scientists. There is a common presumption that mathematical sciences research can be done in a vacuum—that is, that mathematical scientists tend to learn about a problem, retreat to their offices for several months, and reappear only when they have completed their research. This model is not at all true in applied areas, but many biologists have not been engaged in the iterative give-and-take that melds the complementary skills of mathematical and biological scientists to create an advance that neither could have achieved alone. Similarly, many biologists have not seen the powerful difference between using off-the-shelf formulas or software and using a method that is adapted by an experienced mathematical scientist for a particular application.

The charge to the committee asked for recommendations on how the DOE's applied mathematics program can best support its computational biology aims. One thrust for that program should be the refinement of general-purpose tools whose broad biological utility has already been established. Some knowledge of biological applications is often important for pointing this research in optimally useful directions, but intimate familiarity with specific biological problems may be unnecessary. A good example of this dynamic involves applications of Markov chain Monte Carlo (MCMC) methods in biology. These applications are now sufficiently well established that classes of mathematical problems, such as those governing the convergence properties of Markov chains, can be identified whose solution would almost surely prove relevant to a wide array of biological problems.

> **Recommendation: Funding agencies supporting mathematical research related to the life sciences should support the refinement of general-purpose tools whose broad biological utility has already been established. Such research might require specialized review criteria, particularly when the focus is on tool enhancement rather than breakthrough research.**

The committee believes that most advances in the near future in computational biology at all scales will come from adapting established mathematical tools to biological problems. Biology is complicated, and what is needed is insight about which complications can be ignored and which are essential; it is easier to reach that insight when dealing with well-characterized mathematical tools rather than novel ones that might add complexity. These insights will guide the application of sophisticated, but often familiar, mathematical tools to extract as much information as possible from large data sets. In some happy instances, this process will spawn new mathematics. However, no amount of mathematical sophistication can overcome the intrinsic complexity of biological systems. The key will be to achieve steady improvements in our ability to simplify and approximate these systems without losing their essential characteristics. While this process of reduction will certainly require researchers with a good sense of the power and limitations of relevant mathematical tools, it will predominantly require an intimate knowledge of the living systems that they are attempting to approximate. By working for the most part with well-established mathematical tools, the mathematician and the biologist can focus on what data might be missing or what approaches might not have been tried, in order to make the problem tractable. It should be easier to ascertain which features of the complexity can be neglected or ignored, which are essential, and which approaches can provide the best input for mathematical analysis.

The range of mathematical sciences methods that have successfully contributed to biology is very large, as indicated in the rest of this report. Therefore, recommending that the DOE applied mathematics program cover those demonstrated areas of mathematics is not a restriction; in fact, it would require a substantial enlargement of that program's traditional scope. Some of the most promising areas are discussed in the chapter "Crosscutting Themes," but these should be seen as illustrative, not exclusive. As biology itself proceeds, the range of applicable mathematical methods might well expand. Openness, or inclusiveness, will be important to ensure that the methods of mathematics can contribute most effectively to biology.

The federal agencies have set up processes recently to be more responsive to tool development, to the more general aspects of infrastructure support, to the provision of new methods, and to the development of new instruments, new approaches, or software, along with the more traditional forms of infrastructure such as equipment. The agencies have also provided some funding to support what is called discovery science: data mining or exploratory work aimed at gaining a novel insight rather than testing a specific hypothesis. Interdisciplinary research, in general, often requires review processes carefully constructed to permit effective evaluation of novel approaches. More specifically, the plans for generalized tool development will need similar careful review and a mandate provided through the call for proposals.

Recommendation: Funding agencies supporting mathematical research related to the life sciences should give priority to re-

search that addresses intrinsic characteristics of biological systems that reappear at many levels of biological organization: high dimensionality, heterogeneity, robustness, and the existence of multiple spatial and temporal scales.

The committee attempted to identify subdisciplines of mathematics in which broadly based advances would be particularly likely to enhance biological research. However, it concluded that since critical advances had come from nearly every subdiscipline within the mathematical sciences, any such prognostication would be mere guesswork. The committee believes that excellent biology research can be achieved only by answering key questions within that discipline. Specifying a priori the tools to be developed inverts that goal. However, it is clear that if DOE's applied mathematics program is to contribute to computational biology, it should focus on research that is linked to the intrinsic characteristics of biological systems that reappear at many levels of biological organization: high dimensionality, heterogeneity, robustness, and the existence of multiple spatial and temporal scales. All areas of biology will benefit from improved mathematical representations of biological systems.

STRUCTURE OF THIS REPORT

Future biologists will use an enormous variety of mathematical tools. What will be distinctive about their research are the problems they aspire to solve rather than the tools they use to solve them. For this reason, this report is organized primarily around biological, rather than mathematical, themes. Its survey of mathematical challenges in biology, which ranges from molecular to ecological levels of organization, is necessarily cursory. However, the report provides an introduction to the diverse challenges that characterize contemporary applications of mathematics to biology. The daunting task facing policy makers will be to develop mechanisms that encourage the deep integration of mathematics and biology needed for sustained progress across this vast, exciting, and rapidly evolving scientific frontier.

REFERENCES

Alizadeh, F., R.M. Karp, D.K. Weisser, and G. Zweig. 1995. Physical mapping of chromosomes using unique probes. *J. Comput. Biol.* 2: 159-184.
Benzer, S. 1959. On the topology of the genetic fine structure. *Proc. Natl. Acad. Sci. U.S.A.* 45: 1607-1620.
Lipan, O., and W.H. Wong. 2005. The use of oscillatory signals in the study of genetic networks. *Proc. Natl. Acad. Sciences U.S.A.* 10.1073.
Mayr, E. 1982. *The Growth of Biological Thought.* Cambridge, Mass.: Belknap Press.

Naeem, S., M. Loreau, and P. Inchausti. 2002. Biodiversity and ecosystem functioning: The emergence of a synthetic ecological framework. Pp. 3-11 in *Biodiversity and Ecosystem Functioning*, M. Loreau, S. Naeem, and P. Inchausti, eds. New York: Springer.

Palumbi, S.R., S.D. Gaines, H. Leslie, and R.R. Warner. 2003. New wave: High-tech tools to help marine reserve research. *Front. Ecol. Environ.* 1(2): 73-79.

Patil, G.P., and C. Taillie. 2003. Geographic and network surveillance via scan statistics for critical area detection. *Statist. Sci.* 18(4): 457-465.

Post, D.E., and R.P. Kendall. 2004. Software project management and quality engineering practices for complex, coupled multiphysics, massively parallel computational simulations: Lessons learned from ASCI. *Int. J. High Perform. Comput. Applic.* 18(4): 399-416.

Running, S.W., R.R. Nemani, F.A. Heinsch, M. Zhao, M. Reeves, and H. Hashimoto. 2004. A continuous satellite-derived measure of global terrestrial primary production. *Bioscience* 6: 547-560.

Shimizu-Sato, S., E. Huq, J.M. Tepperman, and P.H. Quail. 2002. A light switchable gene promoter system. *Nat. Biotechnol.* 20(10): 1041-1044.

Turner, D.P., S.V. Ollinger, and J.S. Kimball. 2004. Integrating remote sensing and ecosystem process models for landscape- to regional-scale analysis of the carbon cycle. *Bioscience* 6: 573-584.

Zeidler, M.P., C. Tan, Y. Bellaiche, S. Cherry, S. Hader, U. Gayko, and N. Perrimon. 2004. Temperature-sensitive control of protein activity by conditionally splicing inteins. *Nat. Biotechnol.* 22(7): 871-876.

2

Historical Successes

To set the stage for recommendations dealing with tomorrow's productive interaction between the mathematical sciences and biology, this chapter briefly describes some successful interactions of the past. While it is common to hear that biology is only now becoming mathematical, in fact there has been overlap between the two fields for a long time (Cohen, 2004). What is different now is that biology routinely relies on methods from the mathematical sciences to assay and manipulate data and that computational methods have now become powerful enough to model the complexity of biological entities and systems.

THE BEGINNINGS OF POPULATION BIOLOGY

R.A. Fisher, a mathematician trained at Cambridge University, became interested in biology at a crucial time. At the beginning of the 20th century, biologists had rediscovered Mendel's work, and one of their main challenges was to reconcile it with Darwin's theory of evolution. Fisher is one of the scientists credited with ushering in the new era that merged genetics and evolution, sometimes referred to as neo-Darwinian theory, through his work that helped establish the field of population biology. He published several papers on the topic, and his 1930 book *The Genetical Theory of Natural Selection* stands as a landmark of that era. His work demonstrated that statistics is a natural tool for modeling populations.

It is equally interesting to consider how biological data led Fisher to revolutionize the field of statistics. He joined the Rothamsted Experimental Station to apply statistical methods to the mass of data that had been accumulated over many years on field trials of crops. He found that the tools available were inadequate to the task. One review of his work (Yates and Mather, 1963) describes his contribution:

> While at Rothamsted not only did he recast the whole theoretical basis of mathematical statistics, he also developed the modern techniques of the design and analysis of experiments, and was prolific in devising methods to deal with the many and varied problems with which he was confronted by research workers at Rothamsted and elsewhere.

His 1925 book *Statistical Methods for Research Workers*, which introduced analysis of variance (ANOVA) methods to statistics, was a revolutionary advance. "Fisher had by that time also established a rigorous framework for maximum-likelihood methods, which continue to play a central role in statistical inference," according to Aldrich (1997). In 1935 Fisher published *The Design of Experiments*, which was the first book devoted to that subject. Fisher had a significant impact on both biology and mathematical statistics, and his contributions affected the theory and practice of both.

INFERENCE OF GENE FUNCTION BY HOMOLOGY

In the modern world of biology, where sequences of entire genomes are available and the number of such sequences is growing rapidly, one sees the enormous importance of mathematical and computer science methods in advancing biological knowledge. Algorithms are essential at many stages, from finding overlaps of short, noisy sequence strings, to assembling them into complete chromosomes, and to identifying regions that are likely to code for proteins or carry out other genetic functions.

One of the most important tasks is the inference of a protein's function. There are close to 1 million different known and predicted proteins in living organisms. Two proteins are said to be homologous if their similarity is due to common ancestry—that is, if they were generated from the same gene in the genome of an ancestral species at one time in the evolutionary past and their sequences have been sufficiently conserved since that time so that they are still recognizably similar. The number of proteins that have had their functions determined experimentally is, at most, in the tens of thousands, meaning that the functions of over 90 percent of all the proteins in our databases are inferred from homology. In some cases this is easy to do. For example, if one protein has its function determined experimentally and another protein is discovered with a nearly identical sequence, then it is an easy, and quite reliable, extrapolation to assign the same function to the new protein. But, if the sequences of two proteins differ substantially, it is less clear whether they are really homologous to each other. If they are homologous, they are likely to have the same, or a closely related, function, although there are exceptions. Inferring that two proteins are homologous when they are far from identical in amino acid sequence and locating their related sequences in a complete genomic sequence requires the application of mathematical and computational methods that have been developed over the last 40 years.

In the 1960s, Emile Zuckerkandl and Linus Pauling (1965) first realized that DNA and protein sequences, molecules they called "semantides" (for information-carrying polymers) contain the history of their divergence from their ancestors. From the information in genetic sequences, one could, they argued, do "paleogenetics" to find the relationships between genes and therefore also between species. This became the field of molecular evolution, which has flourished and become ever more mathematical as the sophistication of the models for evolutionary change has increased and more complex algorithms have become common for inferring evolutionary events and phylogenetic trees. Also in the 1960s, Motoo Kimura (1968) introduced the theory of neutral evolution and its large contribution to sequence divergence. One effect of his work was simply to emphasize the enormous amount of change that can be observed in biological sequences, which makes paleogenetics that much more challenging, because it means large differences in sequence can accumulate without changes in function. Margaret Dayhoff (1965) produced the *First Atlas of Protein Sequence and Structure*. Among other things, the atlas allowed her to analyze the substitutions observed in closely related proteins and obtain an empirically derived estimate for the rate of substitution of one amino acid for any other. The resulting percentage accepted mutations (PAM) matrices were a much improved measure of the similarity between protein sequences.

To identify the changes between two proteins, one has to find the correct, or at least an optimal, alignment between them. If they are very different, it is not easy to obtain the optimal alignment. Methods referred to generally as dynamic programming, developed by Richard Bellman in 1953, can obtain optimal alignments in such cases very efficiently. Needleman and Wunsch (1970) first published a dynamic programming algorithm to find the optimum alignment between two biological sequences, and over the next several years several variations of that method were developed, differing in how the alignments were scored and how they treated gaps. Most of the efforts were directed at global alignments, where both sequences are aligned along their entire lengths. The more challenging problem was to find local alignments, where only a portion of the two sequences has significant similarity. Local alignment is needed to compare genomes with each other, or even to ask whether a homologue of a particular gene occurs within a genome sequence. Smith and

Waterman (1981) solved the problem of identifying the local alignment using dynamic programming in a way that allows for full use of similarity matrices such as PAM, treats gaps in an intelligent way, and is guaranteed to find the optimal solution efficiently.

While the Smith-Waterman algorithm solves the problem of optimal local alignments, it is not efficient enough for the very large database searches that were becoming necessary by the mid- to late 1980s owing to an exponential increase in database size. The BLAST program, published in 1990, was a major breakthrough (Altschul et al., 1990). This was an important collaboration between two computer scientists (Myers and Miller), a mathematician (Altschul), a medical doctor (Lipman), and a biologist (Gish). It employed a fast heuristic search algorithm for the local alignment problem, and the algorithm's sensitivities are not much reduced from those of the Smith-Waterman algorithm. At the same time, Altschul and Karlin, another mathematician, developed statistical methods to allow computing the significance of the matches found by BLAST (Karlin and Altschul, 1990). When the National Institutes of Health (NIH) made the program available over the newly arrived Internet, biologists around the world suddenly had access to sophisticated database searches to compare their own sequences with the known sequences. Just as the large-scale genome sequencing projects were being contemplated, but before they had truly begun in earnest, this critical piece of software had been developed that would greatly expedite the projects.

EVOLUTIONARY PROCESSES IN POPULATIONS

In the early 1980s population genetics theory took a dramatic turn. Before that time, most theoretical work was focused on the analysis of allele frequencies for two, or perhaps a few, variants in just one or two genes. Interest focused on the frequency spectrum or heterozygosity that one observed at enzyme loci, using protein gel electrophoresis to assay variation. The theory was based primarily on diffusion approximations of two-allele systems (higher-dimensional systems being intractable). When it became clear that surveys of DNA polymorphism would become available via resequencing, it was obvious that some different quantities would need to be studied. With this new kind of data, one would know not only the number of alleles and their frequencies in a sample but also, from long stretches of linked sites, the number of mutational steps by which all the sequences differed from each other. This opened a new window on the evolutionary processes occurring in populations.

To understand the variation revealed by sequencing of alleles, several investigators at that time began focusing on the distributional properties of gene trees. Gene trees, under the standard finite-population-size models, are random structures. The early work, spearheaded by Kingman (1982) and Tajima (1983), showed that many properties of sequence variation could be simply understood in terms of the properties of the genealogical tree relating the sampled sequences. Many other players quickly entered the arena. This area of population genetics became known as coalescent theory. The early theory dealt with the simplest models, which had constant population size, no spatial structure, no recombination, and no selection. Over the next 15 years the theory was generalized to cover models in which all of these limitations had been removed. It is now routine to think about genetic variation in terms of the size and shape of gene trees that relate sampled sequences. It is also routine to simulate samples under many models using efficient algorithms based on the coalescent approach. The models of the coalescent process are often very simple to describe, consisting of relatively straightforward Markov chains, but the genealogical structures that arise are in some cases surprisingly challenging to analyze and rich in their connections with other areas of stochastic processes. For example, coalescent methods for models with selection have connections with the theory of interacting particles and dual processes (Krone and Neuhauser, 1997).

The coalescent approach has led to new insights about the models, to new analytic results (Krone and Neuhauser, 1997), to numerical methods for obtaining likelihoods (Kuhner et al., 2000; Griffiths and Tavaré, 1995), and to very efficient simulation algorithms (Hudson, 1983).

MODELING

As a further illustration of successful interactions between mathematics and biology, consider two separate historical examples of mathematical modeling. The development of the Hodgkin-Huxley equations (Hodgkin and Huxley, 1952) to describe the evolution of action potentials was of profound importance. Their description of ionic currents through ion selective channels provided a paradigm that is still used extensively today in models of cellular electrophysiology. The understanding of excitability that came from their model is also of remarkably general applicability. Perhaps more significant, however, was the recognition that spatially extended systems of excitable components could support waves of invariant form and allow for robust signaling over great distances.

It is now understood that this combination of excitability over spatially extended networks provides the basis for communication and control of many fundamental biological processes. Communication along one-dimensional excitable pathways is remarkably robust and reliable. Yet, in two- and three-dimensional spatially extended networks, other robust patterns (e.g., re-entrant spirals) can arise that overrun the normal function and lead to serious pathology. As a consequence, there is currently a significant effort to understand how to prevent these other naturally occurring, but pathological, patterns and how to get rid of them when they do occur.

A second illustrative example of modeling success is the suggestion of the existence of dendro-dendritic synapses in the olfactory bulb by Rall and Shepherd (1968). It was widely believed, before that time, that axons were excitable, dendrites were passive, and synaptic connections were made between axons and dendrites only. In order to match certain extracellular potentials that were measured experimentally in the olfactory bulb, Rall and Shepherd found that a compartment model with active dendrites was required. As a result, they suggested that dendro-dendritic pathways were likely to exist, and that they provided a novel mechanism for recurrent inhibition. It was only years later that experimental technique improved to the point where such synapses were indeed found to exist. This is an example of the healthy interplay between modeling and experiment, wherein each drives the other in the pursuit of more complete understanding.

MEDICAL AND BIOLOGICAL IMAGING

The last few decades have seen dramatic advances in imaging technology. In medicine, magnetic resonance imaging (MRI) and computed x-ray tomography (CT) are playing increasingly important roles in both diagnosis and treatment, with new applications emerging every year. The importance of the mathematical sciences to biomedical imaging was emphasized in the 1996 National Research Council report *Mathematics and Physics of Emerging Biomedical Imaging*:

> While exponential improvements in computing power have contributed to the development of today's biomedical imaging capabilities, computing power alone does not account for the dramatic expansion of the field, nor will future improvements in computer hardware be a sufficient springboard to enable the development of the biomedical imaging tools described in this report. That development will require continued research in physics and the mathematical sciences, fields that have contributed greatly to biomedical imaging and will continue to do so. (p. 9)

For example, the mathematical foundations for image reconstruction in x-ray CT date back to the work of Johann Radon in the early 1900s. It was around 1970, however, that machines first provided images of value in medical diagnosis, mainly owing to the efforts of A.M. Cormack and G.N. Hounsfield. They observed that by measuring the net attenuation along large numbers of individual x-ray pencil beams, one could reconstruct the attenuation coefficient point by point across a complete cross section of the human body. Nevertheless, several hours of computation time were required to obtain a single image, and the quality was relatively poor. The original numerical methods for reconstruction were based on iterative relaxation of a system of equations, with each equation representing the discretization of an integral measuring the net attenuation along a single line. When Shepp and Logan (1974) and others introduced filtered back projection, it was possible to substantially improve both image quality and reconstruction time, and CT scans became much more practical.

For the purposes of this chapter, it is worth highlighting a very specific mathematical contribution and discussing its ramifications in the context of computational biology more broadly. That contribution was the introduction of the mathematical "phantom" by Shepp and Logan (1974) and Shepp and Kruskal (1978). Consider the situation where one is seeking to compare the performance of a variety of reconstruction methods. The standard approach before Shepp and Logan's work had been to create actual physical models with known characteristics, from which data were measured and reconstruction performed. This seems natural, but errors may stem from inaccuracies in performing the physical experiment as well as in the reconstruction. The Shepp-Logan phantom is a mathematically defined function from which exact (artificial) data can be created, including any desired noise model. Its importance was made clear when Shepp and Logan turned their attention to a ring of high density slightly inside the skull that was observed when the first CT machine was introduced and that was believed to be a previously unrecognized anatomic feature. The use of mathematical phantoms was instrumental in showing that this ring was in fact an artifact of the reconstruction algorithm.

SUMMARY

This chapter gives some indication—but certainly not an exhaustive account—of the long history of interaction between the mathematical and biological sciences. It also demonstrates, by example, the breadth of that interaction: Many areas of biology have been affected by many areas of mathematical science, and the challenges of biology have also prompted advances of importance to the mathematical sciences themselves. Sometimes the benefits of mathematical sciences research have been direct, and sometimes they have arisen in ways that were not predicted. As these examples show, the right mathematical approach can have a dramatic impact on whether or not a particular biological construct is feasible—for example, in the case of finding protein homologues or reconstructing CT scans—and developing the right mathematical representation of a phenomenon can enable very productive research—for example, in the study of populations or exploring signaling mechanisms.

REFERENCES

Aldrich, J. 1997. R.A. Fisher and the making of maximum likelihood 1912-1922. *Statist. Sci.* 12: 162-176.

Altschul, S.F., W. Gish, W. Miller, E.W. Myers, and D.J. Lipman. 1990. Basic local alignment search tool. *J. Mol. Biol.* 215(3): 403-410.

Cohen, J.E. 2004. Mathematics is biology's next microscope, only better; biology is mathematics' next physics, only better. *PLoS Biol.* 2(12): e439.

Dayhoff, M.O., R.V. Eck, M.A. Chang, and M.R. Sochard. 1965. *Atlas of Protein Sequence and Structure.* Silver Spring, Md.: National Biomedical Research Foundation.

Fisher, R.A. 1935. *The Design of Experiments.* Edinburgh: Oliver and Boyd.

Fisher, R.A. 1930. *The Genetical Theory of Natural Selection.* Oxford: Clarendon Press.

Fisher, R.A. 1925. *Statistical Methods for Research Workers.* Edinburgh: Oliver and Boyd.

Griffiths, R.C., and S. Tavaré. 1995. Unrooted genealogical tree probabilities in the infinitely-many-sites model. *Math. Biosci.* 127(1): 77-98.

Hodgkin, A.L., and A.F. Huxley. 1952. A quantitative description of membrane current and its application to conduction and excitation in nerve. *J. Physiol.* 117: 500-544.

Hudson, R.R. 1983. Properties of a neutral allele model with intragenic recombination. *Theor. Popul. Biol.* 23(2): 183-201.

Karlin, S., and S.F. Altschul. 1990. Methods for assessing the statistical significance of molecular sequence features by using general scoring schemes. *Proc. Natl. Acad. Sci. U.S.A.* 87(6): 2264-2268.

Kimura, M. 1968. Evolutionary rate at the molecular level. *Nature* 217(129): 624-626.

Kingman, J.F.C. 1982. On the genealogy of large populations. *J. Appl. Prob.* 19A: 27-43.

Krone, S.M., and C. Neuhauser. 1997. Ancestral processes with selection. *Theor. Popul. Biol.* 51(3): 210-237.

Kuhner, M.K., J. Yamato, and J. Felsenstein. 2000. Maximum likelihood estimation of recombination rates from population data. *Genetics* 156(3): 1393-1401.

National Research Council. 1996. *Mathematics and Physics of Emerging Biomedical Imaging.* Washington, D.C.: National Academy Press.

Needleman, S.B., and C.D. Wunsch. 1970. A general method applicable to the search for similarities in the amino acid sequence of two proteins. *J. Mol. Biol.* 48(3): 443-453.

Rall, W., and G.M. Shepherd. 1968. Theoretical reconstruction of field potentials and dendrodentritic synapse interactions in olfactory bulb. *J. Neurophysiol.* 31: 884-915.

Shepp, L.A., and J.B. Kruskal. 1978. Computerized tomography: The new medical x-ray technology. *Am. Math. Mon.* XX: 420-439.

Shepp, L.A., and B.F. Logan. 1974. The Fourier reconstruction of a head section. *IEEE Trans. Nucl. Sci.* 21: 21-43.

Smith, T.F., and M.S. Waterman. 1981. Identification of common molecular subsequences. *J. Mol. Biol.* 147(1):195-197.

Tajima, F. 1983. Evolutionary relationships of DNA sequences in finite populations. *Genetics* 105: 437-460.

Yates, F., and K. Mather. 1963. Ronald Aylmer Fisher 1890-1962. Pp. 91-120 in *Biographical Memoirs of Fellows of the Royal Society of London*, Vol. 9. London, England: The Royal Society.

Zuckerkandl, E., and L. Pauling. 1965. Molecules as documents of evolutionary history. *J. Theor. Biol.* 8: 357-366.

3

Understanding Molecules

INTRODUCTION

Cells are composed of molecules, and their properties are largely determined by the chemical reactions the molecules perform and by the structures the chemicals form. The molecules range from the small and ubiquitous, such as water and various salts and metal ions, to the very large molecules that are specific to living systems, such as the genetic material of DNA. In between are a wide variety of organic molecules necessary for life, including sugars and lipids, vitamins and other enzyme cofactors, and the nucleic acid and amino acid monomers required for DNA, RNA, and protein synthesis. The molecules interact in many ways and are capable of recognizing one another; some are active in the form of larger complexes. These dynamic interactions are the essence of the processes in living cells. Cells in turn have mechanisms for accurately controlling their composition, for obtaining from the environment molecules they can use, and for synthesizing those that they need. They can modify their composition depending on their environment and their fate. At every cell division, they must essentially duplicate every component, except that occasionally cells undergo asymmetric division so that the two daughter cells can take on different roles.

For the last 50 years, the field of molecular biology has examined the hereditary and information-carrying molecules of DNA, RNA, and proteins. These molecules will be the focus of this chapter. Determining the double-stranded structure of DNA made clear how genetic information is replicated and passed from generation to generation. Since then, much effort has been devoted to constructing a flow chart for the various forms of molecular information. One of the fruits of this labor has been a sense that many of the main information highways are known—for example, the central dogma that describes the irreversible flow of information from DNA to RNA to proteins. The fact that reality is more complex and that surprises continually emerge, such as the recent appreciation for the variety of roles played by RNA, shows that this enormously successful field still has a long road ahead before it will be "solved" in any comprehensive sense.

One of the most successful achievements of molecular biology has been a nearly complete catalogue of underlying DNA codes for a diverse set of information-bearing macromolecules: The genomes of humans and many microbes, plants, and other animals are known with considerable completeness and accuracy. There has thus emerged the feeling that biological explanation is no longer primarily to be sought in finding new molecular actors but in understanding their individual functions and the patterns of organization and interaction that collectively determine the functions of the cell.

THE MATHEMATICS-BIOLOGY CONNECTION

Molecular biology has always relied heavily on mathematics. From the analysis of sequences to techniques for determining the three-dimensional structures of molecules to studies of the dynamics of entities ranging from individual molecules up to entire networks, mathematical techniques and computational algorithms are critical.

Because of the rapid advances in the technology for DNA sequencing, DNA sequences are now easily obtained, and protein sequences can be inferred with reasonably high accuracy and completeness. Thus, we now have an abundance of those sequences for analysis. DNA, RNA, and proteins are all linear polymers, or strings, made from a small alphabet of residues, 4 for DNA and RNA and 20 for proteins. The specific properties of any molecule, or the functions it serves, are determined by its sequence and its structure (in the appropriate context), though of course the structure is a result of the sequence and the molecule's environment. One of the mathematical challenges of biology, then, is determining the mapping from sequence space to function space. The set of linear sequences over a small alphabet leads quite naturally to the concept of sequence space and the universe of all possible sequences. The concept of function space is a little harder to imagine, but certainly we could categorize all of the functions we know and consider them to be a partial set of all possible functions.

For proteins, and for some RNAs, function is critically dependent on structure, and knowledge of structure can provide information about function. Hence, the mapping from sequence space to structure space is part of the challenge, and may be part of the solution, of creating a map from sequence space to function space. The structure of a protein provides strong clues about its biochemical function—for example, the mechanism for action by an enzyme—but at the moment, there have been only a few successes in predicting biological function from sequence. The structures of these macromolecules are also important for other research purposes—for example, they are the starting point for predicting biochemical action or for modeling the dynamics of the macromolecules, for suggesting ways to inhibit the action of undesired proteins, for predicting potential chemical inhibitors or activators of a given protein, or for altering a protein's functionality through its environment or through reengineering its sequence and, consequently, its structure. Therefore, developing the ability to map from sequence space to structure space is a critical challenge that, if met, would have a significant impact on all biological sciences and on our understanding of life. Currently, inferences about structure and function rely on the simple assumption that sequences that are "close" in sequence space (using metrics determined from studies of evolution) are likely to map to nearby points in structure and function space. That is generally true, but there are complications. Short stretches of a protein can be exceptions to this general situation, and larger proteins are composites of folded segments or domains. In the absence of experimental determination through an x-ray crystallographic structure determination, we do not know in any detail how to ascertain where the boundaries are for domains. We also do not know which sequence differences are most critical or might be most indicative of exceptions or might most effectively predict the biological function.

Even a catalog of all the components of a cell (a complete "parts list") detailing not only their sequences but also their structure and function would not really explain the properties of that cell, because the system is far from equilibrium and in a very dynamic state. The properties of molecules often depend on their dynamics, from the catalytic activities of enzymes to the assembly of multicomponent structures, and many of a cell's molecules need to be transported to specific locations within or outside the cell in order to perform their functions. Cells sense their environment and respond to various stimuli by sending signals throughout the cell and to neighboring cells, modifying their behavior. Metabolic networks are subject to feedback regulation and other kinds of control, and the expression of specific genes is controlled by networks of regulatory factors and their interactions with each other and with cellular signals. Because many cellular processes are due to the actions of a small number of molecules, stochastic fluctuations cannot be ignored. In general, then, understanding the properties of cells depends on modeling the dynamics of the individual molecules and their interactions. The dynamics has both spatial and temporal components and is largely nonlinear and discrete. Mathematical analysis of dynamical systems has been essential for our current understanding, but the field appears to be on the verge of a major expansion. New technologies offer the opportunity to greatly increase what we can measure about cellular components. Using those data to inform dynamical models of the cell is critical to advancing our understanding of biology and is likely to tax existing mathematical techniques, opening up new areas of mathematical and biological research.

AREAS OF MATHEMATICAL APPLICATIONS FOR MOLECULES

Sequence Analysis

The central role of sequences in mathematics is unquestioned. The discovery that DNA, RNA, and proteins are all composed of linear sequences provided a strong link between active areas of research in both molecular biology and the mathematical sciences. A well-developed set of results for sequence analysis has been developed in computer science, some of which are useful for biological problems (Gusfield, 1997; Waterman, 1995). The most classical problem of string searches is to find exact occurrences of a sequence (usually short, which we will call the "pattern") in another sequence (usually long, called the "text"). The Boyer-Moore and Knuth-Morris-Pratt algorithms are a sophisticated pair of algorithms that have been developed to alleviate this problem (Pevzner, 2000). These techniques make it possible to find exact matches to the pattern in linear time (time that is proportional to the *sum* of pattern and text lengths) or better; the obvious naive method takes quadratic time (time proportional to the *product* of pattern and text lengths). Suffix trees also accomplish this task and, more importantly, are also useful for more general problems (Gusfield, 1997). Certain biological problems involve exact patterns, an important example being restriction sites (usually four to eight base pairs in length), where certain enzymes cut DNA molecules.

However, it is usually the case that sequence problems in biology involve approximate matching. Sites where a protein can bind to DNA, for example, usually are very inexact, in the sense that there may be a few distinct positions where certain bases are strongly preferred but no unique binding site. Dynamic programming (DP) is a useful method for many problems of approximate pattern matching (Waterman, 1995). Usually taking quadratic time, various DP algorithms can find the approximate location of patterns in texts, the best relationship between two and more sequences (the alignment problem), and the best overlap between two sequences. These algorithms are indispensable for sequence assembly, the process of inferring a genome sequence from an oversampled set of short, randomly located sequences. One alignment problem is to find the best relationship between the letters of two sequences, and another is to find the best segments that match well. The latter problem, usually called local alignment, is most often used in database searches, where one wants to find domains of one sequence that are similar to domains of another sequence by common ancestry—that is, domains that are homologous. Because the databases of known sequences have become very large, DP is too slow for the full searches. Various heuristic methods, such as BLAST (Altschul et al., 1990), have been used for this problem and can be fairly effective, but greater sensitivity at detecting distant homologies is needed.

Critical to the assessment of similarity between sequences is a model for evolutionary processes. Before protein and DNA sequences, the information used to classify an organism and infer its history was its observable properties (e.g., wings or gills) as well as fossil evidence. But once it was realized that DNA and protein sequences contain traces of their history, new avenues to studying evolution—paleogenetics—were opened up (Zuckerkandl and Pauling, 1965). From alignments of clearly homologous sequences, it was possible to determine empirically the rates of residue substitution. Those rates allow optimum alignments to be determined over much larger evolutionary distances and inferences of homology to be much more widely applied. From those alignments it is then possible to determine the evolutionary relationship of molecules, and the species that contain them, at a much higher resolution than previously possible.

This sort of information is commonly represented by phylogenetic trees (Felsenstein, 2004). The structure of these trees is generally inferred by one of three approaches: parsimony, distance matrices, or likelihood. Parsimony uses the principle of "least evolution" to estimate which sequences are most closely related. Finding the most parsimonious tree is an NP-hard problem, but many heuristic methods have been devised to approach it. When organisms are sufficiently closely related, parsimony is a reasonable model. Otherwise, multiple changes may have occurred, and it can be quite misleading. For the distance matrix approach, a distance is defined between each pair of sequences based, for instance, on pairwise sequence alignment scores from DP or from a position-by-position score relating a pair of sequences sampled from a full multiple alignment. Can a tree generated this way have the property that the distance between any two sequences is the same as the sum of distances through the vertices that connect those sequences? There is some elegant work on this problem, which includes the celebrated four-point condition, but it is almost never the case that a distance matrix is additive. Finally, likelihood models assume a stochastic model for evolution of the positions along branches of the tree. Likelihoods for two trees can be compared using classical likelihood ratios. Modern Bayesian methods, including simulated annealing, and Markov chain Monte Carlo methods (MCMCs) are employed. Work in this area is quite active, and the approach is considered by many to be the method of choice.

All of these methods have significant limitations. The number of possible trees is unmanageable for any reasonable number of sequences. Also, the methods all depend on the calculated alignment of sequences that contain uncertainties that are not all accounted for in the tree-building methods. Finally, simplifying assumptions about the independence of the positions and the uniformity of substitution rates limit the resolution that can be obtained.

Probabilistic models, such as hidden Markov models (HMMs), have recently made significant contributions to biological sequence analysis (Durbin et al., 1998). In protein comparisons, these approaches allow for position-specific variability in substitution scores. In modeling interacting domains of DNA and RNA, they allow for a more biophysical treatment of the interaction. And in ab initio predictions of gene sequences, they can better capture the statistical characteristics of genes, including both content features and signals that delineate boundaries between different segments. For protein-coding genes, such methods can be reasonably effective, but for genes that code for RNAs, the problem is much more difficult and challenging.

Structure Analysis

Determining the three-dimensional structure of macromolecules from experimental data, such as x-ray diffraction patterns and nuclear magnetic resonance measurements, is a mathematically demanding task. A typical protein structure must be inferred from enormous amounts of information that indirectly reveal the relative locations of key atoms in a biomolecular structure. The mapping from structure to data is straightforward, but the inverse problem, going from data to structure, is quite complex. The task is made harder because the experimental data are usually still too sparse to make a unique inversion possible. Instead, models must be built to overcome the intrinsic experimental ambiguity. This is done, for instance, by forcing the models to conform to the best current knowledge of molecular forces and/or to adhere to a presumed structural component, such as a peptide backbone in protein chains. Even so, constructing these models leads to demanding optimization problems.

Because one-dimensional sequence data abound but three-dimensional structures are generally more useful, understanding the mapping from sequence to structure is a major goal of molecular biology. Currently, the most reliable methods for predicting the structure of a new protein are based on homology. If the new protein sequence can be inferred to be homologous to a protein with known structure, then it can also be reliably inferred that the structures of the two proteins are quite similar. In general, protein structures are more highly conserved through evolution than their sequences, so the challenge is to identify distantly related homologous proteins, and progress is ongoing in this direction, as described in the last section and Chapter 2. Despite decades of research, there has been only modest progress in predicting protein structures from their sequence alone, ab initio. Experiments have shown that at least for moderate-sized proteins, sequence information alone is sufficient to specify the molecule's three-dimensional shape. Understanding and then reproducing in the computer this transformation of one-dimensional information to three-dimensional information has come to be known as the protein-folding problem. The development of the modeling and computing capability needed to enable calculating structure from sequence data would clarify many basic issues about protein structure and function. This capability would also be of practical use in the emerging field of protein engineering.

Considerable progress has been made on the protein-folding problem using, on the theoretical side, the statistical mechanics of energy landscapes and, on the experimental side, knowledge of protein engineering and physicochemical kinetics. Of course some progress can simply be attributed to the rapid increase in the power of large-scale computing. The theoretical advances come from understanding how to characterize the universal topological properties of the energy surfaces of molecules; this challenge is deeply connected to the problem of characterizing the computational difficulty of optimization problems. Since the proteins fold on their own, and fairly rapidly given the enormous number of possible folding paths and the NP-completeness of the optimization problem, protein folding must somehow be constrained. The energy landscapes of real proteins must be rather smooth, resembling a funnel leading to the three-dimensional folded structure rather than the very rugged energy landscapes that could be present in principle. The rugged energy landscapes have globally different optima very far apart in structure, and this is a handicap to any search for optima.

Until recently, the energy functions used to simulate the folding process, built up from the interactions of fragments of protein, were too rugged to yield correctly folded structures. Instead, depending on the fine details of the optimization process, many unrelated structures were predicted to represent the global free energy minimum. The funnel-landscape idea offers a mathematical tool for developing more accurate energy surfaces by using bioinformatic patterns in the known database of protein structures. Algorithms based on such hybrids of bioinformatics and physical energy functions begin to approach the accuracy of structure predictions based on homology. Nonetheless, innovative ideas are still needed to enable predicting the folded structure of proteins.

Useful as the classical protein-folding problem has been in stimulating interest in the physical chemistry of biological macromolecules, it is important to recognize that its "solution," at least across any broad sampling of protein structures, is unlikely. In cells, the self-assembly processes that build up biological structures from their components—including the folding of many proteins—occur in highly specialized environments. For example, it has become clear during the past 15 years that the folding of many proteins is facilitated by local environments provided by proteins known as chaperones.

To predict the structure of RNAs, efficient algorithms based on thermodynamic parameters and DP methods have been available for over 20 years. Yet the best predictions are not very accurate, especially for long RNAs. The difficulty of making accurate predictions is undoubtedly due to some combination of the following facts: There is an enormous number of possible structures with nearly equal predicted energies; available thermodynamic parameters have limited accuracies; many RNAs require cofactors, such as magnesium, in the medium in order to fold properly; and efficient algorithms eliminate some possible structures, such as pseudoknots, that may be necessary to obtain the correct folding. As with proteins, RNAs can be folded most accurately by using information from homologous RNAs with known structure. But in the case of RNAs, it is also very useful to have related RNAs that, presumably, fold into nearly identical structures even if none of the structures are known in advance. Because base-pairing is the strongest force determining RNA structures, finding base-pairing patterns that are consistent for all of the sample sequences can lead to correct structure prediction. Sankoff (1985) published an algorithm for determining the optimum combination of alignment and structure scores for a collection of RNA sequences. However, that algorithm is computationally too expensive to be useful for typical numbers and lengths of sequences, so heuristic approximations have been developed in recent years to solve this problem.

Dynamics

Many experiments show that biomolecules do not exist as the single structures envisioned in the simpler forms of structure reconstruction. Instead, they are present in the body as an ensemble of structures, thermally populating a complex energy landscape. Some reconstructions try to take this ensemble character into account, but it is difficult to do so, and there is no mathematical understanding of the limits of application for these algorithms. The problem is no longer one of carrying out a simple optimization but one of achieving a statistically reliable sampling of structures that conforms to our knowledge of molecular forces. Understanding the dynamics of molecules is not simply an exercise in modeling random fluctuations, because in many cases the dynamics of proteins are essential to their function. It is only by allowing some transformations to occur quickly and other thermodynamically possible transformations to occur slowly that a cell can control its behavior.

Computer simulation has made much progress in illustrating the motions of proteins, yet there are several problems in trying to simulate completely from molecular dynamics how biological molecules function. The first challenge is timescale. While the atoms in proteins carry out the bulk of their motions in less than a picosecond, the fastest characteristic response time for a cell is in the millisecond range. Evolution has tuned and selected molecular properties to deliver results on the latter timescale. Thus, the timescales of importance range over nine orders of magnitude, and one cannot argue, as is done in computational fluid dynamics, that short-timescale behavior is important only at small spatial scales and long-timescale behavior is important only at large spatial scales. The reason for the timescale gap is that functionally important motions are rare: A movie of the atoms in a biomolecule could be mostly a great bore, interrupted by a few dramatic moments. Overcoming the timescale problem will involve understanding the structure of the possible biomolecular motions and developing mathematical ways to survey all the possibilities in a statistically meaningful way.

The high degree of accuracy to which functional dynamics must be computed is also a challenge for direct simulation. For instance, there might be only a small amount of energy difference between one binding state and another, yet the choice of binding state is a critical determinant of which of several signaling cascades takes place. Thus, calculations must be done with great precision. However, because even the smallest biological macromolecules contain thousands of atoms, how can the energy be determined this accurately? Cells manage to achieve reliable specificity, so there must be a principle at work, much like the funnel-landscape idea for folding, that simplifies the problem. But discovering this principle is a challenge. Can it be learned after enough four-dimensional binding data have been determined using machine learning techniques? In his 1905 papers on Brownian motion, Einstein noted that at the large size scale of macromolecular assemblies in the cell, energies of very small order still reign and determine where things go. As an example, it is now believed that the stability of the chromatin packaging of DNA comes from *attractive* electrostatic forces between ions of *like* sign. Understanding the connection of very small forces to the macromolecular interactions occurring within the cell remains a grand computational challenge.

Interactions

Interactions between macromolecules play many essential roles. They occur in all combinations, with protein-protein and protein-DNA/RNA being the most studied. Complexes of multiple proteins are important in metabolic reactions where reactants can pass between enzymes to increase efficiency and reduce concentrations of intermediates. They are important in large intracellular structures, such as various types of filaments that provide both rigidity and directed motion within and between cells. Protein-protein interactions are essential in various signal transduction pathways where proteins modify one another to alter their properties. Cascades of such protein modifications are at the heart of processes that transfer information and signals between cellular components, and they allow cells to respond to their environment by altering their behavior.

Experimental approaches to determining the interactions between proteins have greatly expanded in the last few years, and there is now considerable data about specific interactions. However, these experimental methods tend to produce noisy data, and it remains a challenge to extract the true information. Some computational methods have proven useful in detecting interacting proteins, such as correlated occurrence in phylogenetic comparisons. While not currently feasible, computational methods to predict protein interactions based on compatible surfaces— essentially predicting the docking of large molecules—would be a major advance. One practical application of progress in this area would be the design of small-molecule drugs that interfere with specific protein-protein interactions.

Protein interactions with DNA and RNA are the primary mechanisms for controlling gene expression. In specific cells under specific conditions, only a subset of genes is expressed at any given time. Based on environmental signals, a fraction of the regulatory proteins will be expressed and active. Their sequence-specific interactions with regions of DNA will determine the set of genes expressed during the next interval. Statistical methods for predicting DNA target sites for specific proteins, given previous examples of similar sites, can be useful for discovering new regulatory interactions. What is needed is a recognition code that maps from the protein sequence (and utilizes the known structures of the transcription factor families) to a pattern that describes the family of DNA binding sites. Statistical methods have been applied to this problem, but current performance is far from adequate. Combinations of statistical methods and biophysical principles appear promising but are just in their infancy (Benos et al., 2002).

Many of the interactions between macromolecules define networks, both metabolic and regulatory. Simply knowing the components of the networks and the connections between them is not sufficient to understand or model their behavior. The levels of different components of the networks can vary over space and time and are often small enough that discrete modeling is important. In regulatory networks, each of the standard binary logic gates can be implemented by combinations of small numbers of proteins interacting with DNA; in other cases, continuous output values, sometimes with high gain, can be achieved by a similarly small number of components. Even in the best-studied cells, the network of functional interactions is only now being mapped out at the scale needed. All mathematical descriptions that adequately model the diverse components at suitable resolution while also capturing the stochastic character of intermolecular partnership and the complexity of individual biomolecular dynamics remain a distant goal but one that the community of mathematical and physical scientists is beginning to tackle.

FUTURE DIRECTIONS

There are many daunting mathematical challenges to understanding and modeling the molecular aspects of biology. Technological advances continue to increase the quantity of data that can be obtained and, more importantly, to enable the collection of data that were not previously available. This flood of data promises enormous advances in our understanding of biology in the coming decades, but that promise depends on the coincident development of mathematical approaches and applications.

Sequence databases will continue to grow at a rapid rate (currently doubling about every 15 months), increasing our ability to identify homology relationships and improving the evolutionary models on which those identifications depend. Increasingly comprehensive data will allow us to abandon some of the simplifying assumptions that were previously required and to develop more realistic models. The great increase in sequence data will also step up the demand for improvements in the mapping from sequence to structure and function.

Improved mathematics is also required for advances in the biophysical modeling of molecules and cells. Predictions of structures and interactions based on physical principles should significantly complement the results from evolutionary studies and are necessary for advances in protein design and other areas of synthetic biology and biological engineering. Improvements in modeling the dynamics of biological systems that take into account the range of important scales of space, time, and number of components—for example, whether discrete or continuous models are needed—are necessary to develop accurate and predictive models of cellular behavior.

Mathematical models at all scales are important in biology and will become increasingly so. One of the critical areas of development will be the interactions between models and experiments. Every scientist develops models from data and uses those models to design the next experiments. But as the data become increasingly numerous and complex and the models ever more mathematical, it will become even more necessary to rely on mathematical and computational methods in the design of experiments. Determining the set of plausible models that explain the current information and then identifying the most informative experiments to distinguish between those models is a challenge even now and will be an increasingly central challenge in the future. For example, expression data can provide information about sets of plausible regulatory networks but are unlikely to define a single best network. By comparing the different implications of plausible networks, it should be possible to design experiments that best distinguish between them (Ideker et al., 2000).

The molecular level deals with the most basic components of the cell. Key molecules include the heritable material that permits the propagation of biological properties from generation to generation, while also displaying the variation that underlies evolution. Interacting sets of biological molecules are required for the behavior of cells and organisms, but they are not sufficient to predict this behavior. Cells, which will be discussed in the next chapter, provide a first example of the large gulf between successive levels of biological organization.

REFERENCES

Altschul, S.F., W. Gish, W. Miller, E.W. Myers, and D.J. Lipman. 1990. Basic local alignment search tool. *J. Mol. Biol.* 215(3): 403-410.

Benos, P.V., A.S. Lapedes, and G.D. Stormo. 2002. Is there a code for protein-DNA recognition? Probab(ilistical)ly. *Bioessays* 24(5): 466-475.

Durbin, R., S.R. Eddy, A. Krogh, and G. Mitchison. 1998. *Biological Sequence Analysis: Probabilistic Models of Proteins and Nucleic Acids.* Cambridge, U.K.: Cambridge University Press.

Felsenstein, J. 2004. *Inferring Phylogenies.* Sunderland, Mass.: Sinauer Associates, Inc.

Gusfield, G. 1997. *Algorithms on Strings, Trees and Sequences: Computer Science and Computational Biology.* Cambridge, U.K.: Cambridge University Press.

Ideker, T.E., V. Thorsson, and R.M. Karp. 2000. Discovery of regulatory interactions through perturbation: Inference and experimental design. Pp. 305-316 in *Biocomputing 2000: Proceedings of the Pacific Symposium,* Vol. 5. Singapore: World Scientific.

Pevzner, P. 2000. *Computational Molecular Biology: An Algorithmic Approach.* Cambridge, Mass.: MIT Press.

Sankoff, D. 1985. Simultaneous solution of the RNA folding, alignment and protosequence
    problems. *SIAM J. Appl. Math.* 45: 810-825.
Waterman, M.S. 1995. *Introduction to Computational Biology: Sequences, Maps and Genomes.*
    London, U.K.: Chapman and Hall.
Zuckerkandl, E., and L. Pauling. 1965. Molecules as documents of evolutionary history. *J.
    Theor. Biol.* 8: 357-366.

4

Understanding Cells

INTRODUCTION

To understand cells, one must understand the macromolecular structure of cells, the spatiotemporal patterns and mechanisms of cellular
dynamics, the connections between cellular dynamics and cellular functions, and the connections between cells and higher levels of organization, such as tissues and organs. Understanding cells is intrinsically more
difficult than understanding molecules because there is no cellular coun-
terpart to the linear sequence of nucleotides and amino acids that provides much of the information necessary for predicting the structure and
function of nucleic acids and proteins. Moreover, eukaryotic cells are
highly compartmented and contain both a nuclear genome and one or
more organellar genomes. With rapid increases in computational power
and the sophistication of biophysical measurements, one can imagine
constructing reasonably exact models of the dynamics of DNA and proteins, whereas all quantitative descriptions will still be approximate for
cells. Thus, the main challenges for the mathematical analysis of cells are
not computational but reside instead in the basic challenge of how to
model features of interest. The primary challenge in this area for the next
decade is the systematic formulation of reduced-order representations of
cellular structure and dynamics, drawn from increasingly complex data
and validated in model-driven experiments.

There is a long and successful history of mathematical modeling of
cellular functions. The success stories come from systems that are rich in
data and for which models can be validated or at least put in direct corre-

spondence with experiments. Models of the endocytic cycle, signal transduction cascades, and the cell division cycle have been productively used to organize data, extract "laws" for cellular processes, and even engineer cellular systems. Nevertheless, the models formulated over the past two decades build primarily on population-level measurements with poor spatial and temporal resolution. Without exception, the dynamics predicted by early models were richer than those predicted by the corresponding experimental data. For example, current models of cell cycle dynamics have more variables than can be measured experimentally. This experimental limitation is changing as a result of the rapid development of imaging techniques and high-throughput assays of cellular processes. The result should be an increased ability to evaluate models, which is the limiting step in improving them. It is now possible to collect multivariable and spatiotemporally resolved data on cellular processes ranging from molecular trafficking and signal transduction to integrated responses such as the cell division cycle and cell migration.

In spite of these improvements in experimental capabilities, the quantitative models that emerge will not be as resolved and detailed as the models used in, for example, the aerospace and semiconductor industries. This difference reflects the intrinsic variability of biology, the immense spatiotemporal complexity of cells, and our incomplete knowledge of cellular processes. In many areas of the physical sciences, a coarse model can be very simple and yet fairly accurate, and one only needs to develop heterogeneous, multivariable, spatially resolved models when dealing with second- or higher-order effects. In contrast, even the simplest cellular models must be extracted from heterogeneous, multivariable, spatially resolved experimental data. Learning how to manage these data and mine them to extract computationally manageable models of cellular functions is the key challenge to quantitative understanding of cells.

Exemplification of These Issues

In the early 1980s, Steven Wiley and his colleagues formulated kinetic models of receptor-mediated ligand internalization (Wiley and Cunningham, 1981). The models were initially developed for the epidermal growth factor (EGF) receptor, a key regulator of cell and tissue functions across species (Wiley, 2003). The models described the kinetics of ligand-receptor binding, internalization, recycling, and degradation. The mathematical models were in the form of small systems of ordinary differential equations that were integrated in time and coupled with standard optimization and parameter estimation routines to extract the model parameters (Figure 4.1).

Based on experiments with radioactively labeled EGF ligands, the

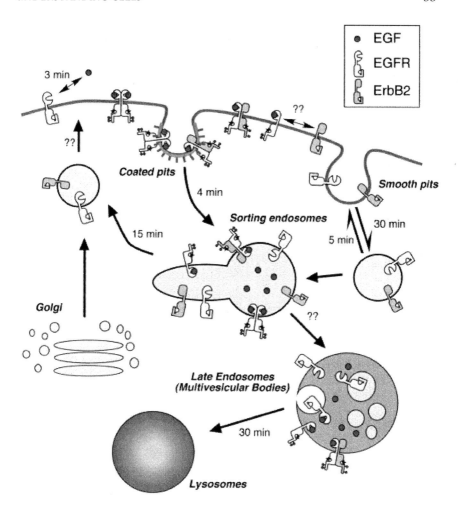

FIGURE 4.1 Trafficking of ErbB receptor family. Only the epidermal growth factor receptor (EGFR) and ErbB2 proteins are shown for clarity, but the behaviors of ErbB3 and ErbB4 are similar to that of ErbB2. Activated EGFR and EGFR:ErbB2 heterodimers are internalized through a coated pit pathway, but other members of the ErbB family are probably internalized by a smooth pit pathway. The numbers next to the arrows represent the approximate mean time of the specific process. The time constants for heterodimerization and formation of multivesicular bodies are unknown. The mean time for lysosomal degradation is a combination of the time necessary for multivesicular body formation and for lysosomal fusion. Reprinted from *Experimental Cell Research*, 284, H.S. Wiley. Trafficking of ErbB receptors and its influence on signaling, pp. 78-88 (2003), with permission from Elsevier.

models could extract the rate constants for different processes, such as endocytic uptake or recycling. Before this modeling effort, the only quantitative measures of ligand-receptor dynamics in cells were related to ligand-receptor interactions. The models of Wiley and colleagues produced new quantitative measures of ligand-receptor dynamics. In this way the biological effects of various ligands could be interpreted in terms of the quantitative differences of a larger number of rate constants for the multiple steps in the endocytic pathway. The model was also critical for suggesting the functional roles of the different parts of the EGF receptor. The EGF receptor is a large protein that combines multiple functions, including ligand binding, receptor phosphorylation, internalization, endocytic sorting, and recycling. By analyzing the ligand uptake data in cells that express mutant receptors and fitting these data to models, the functional roles of specific residues could be identified by changes in the rate constants of specific cellular processes (Wiley et al., 1991).

This modeling and experimental approach has been validated by a number of experiments and used to parse the dynamics of internalization and trafficking for other ligand-receptor systems (Wiley et al., 2003). In the meantime, the experimental tools used to study these processes have changed. It is now possible to visualize multiple steps of ligand-receptor interactions and the endocytic cycle at the single-cell level and in real time (Sorkin et al., 2000). Furthermore, many new molecules have been identified in each of the steps of the endocytic cycle. For example, the recruitment of the ligand-bound receptor to the coated clathrin pits and the transfer of receptors to early endocytic compartments rely on tens of proteins. Protein-protein interactions in this system can be assayed using powerful biophysical techniques, and new components can be discovered by high-throughput proteomic approaches (Blagoev et al., 2003). In connection with this increased appreciation of the underlying molecular complexity of the system, it becomes necessary to rethink the mechanistic meaning of the endocytic rate constants predicted by the original model. How should the current model be changed to incorporate new data? Should the new models necessarily have more variables and parameters? Or, alternatively, can the old models be "parameterized" by new interactions? Given the structural complexity of living cells and the significant cell-to-cell variations, it is unlikely that useful models will account for every protein discovered in the endocytic cycle. But, for this and every other cellular system, it remains an open question how to use the new and much richer data to formulate the simplest model that can be used to correlate data and formulate new experiments. The main point is that, at this time, the data are richer than the models, which was not the case in the 1980s and 1990s.

CELLULAR STRUCTURES

In the same way that knowledge of the composition and structure of biomolecules is the key to understanding their biological function, detailed knowledge of the structure of cells is a prerequisite for the quantitative understanding of cellular functions. Mathematics plays an important role in characterizing the intracellular architecture (Figure 4.2). At one

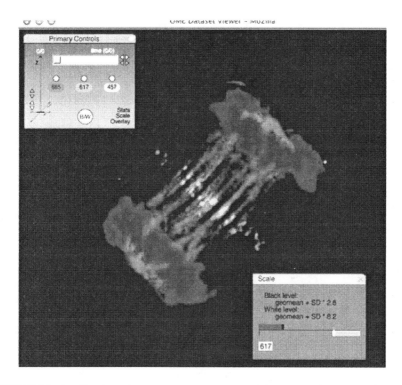

FIGURE 4.2 Applications for quantitative imaging. The image shows an XlK2 cell during the process of cytokinesis stained for DNA, microtubules, and the aurora-B protein kinase. Although the image demonstrates the relative localization of different cellular components and structures, quantitative analysis reveals specific characteristics that can be used to assay the effects of inhibitors on expressed proteins. For example, integrating the signal from a DNA-specific fluorophore might reveal defects in chromosome segregation during mitosis. Measuring the overlap of microtubules and aurora-B (using, for example, a cross-correlation analysis) within a subregion of a dividing cell might be a means of assessing effectors of cytokinesis. The image is displayed within the Open Microscopy Environment Image Viewer. The viewer includes support for displaying multidimensional image data (top left) and some of the associated metadata about each image (bottom right). SOURCE: Swedlow et al., 2003.

level, the mathematical sciences have enabled progress through contributions to the development of instrumentation and other tools. For instance, the explosion of information about the spatiotemporal dynamics in cells would have not have been possible without earlier progress in imaging and data-processing algorithms. Tools such as deconvolution microscopy rely on robust numerical algorithms. Moreover, sophisticated data-processing algorithms have become more accessible to biologists through packages such as Matlab and Metamorph. New data-related challenges are emerging. For example, the assembly of large imaging datasets requires advances in bioinformatics and data mining. The informatic aspects of intracellular imaging are therefore receiving increased attention (Swedlow et al., 2003; Young et al., 2004).

Quantitative imaging enables the formulation of data-driven models of intracellular dynamics and transport. The most notable examples include the quantitative analysis of the dynamics of Golgi to plasma membrane transport (Hirschberg et al., 1998) and nucleocytoplasmic shuttling (Smith et al., 2002). In both cases, green fluorescent protein (GFP)-based imaging provided data of unprecedented spatiotemporal resolution; nevertheless, it was possible to formulate simple compartmental models based on a small number of linear ordinary differential equations. Newer models enable the identification of the rate-limiting steps of the process and the formulation of testable hypotheses. Conclusions about the mechanisms in each case were based on the analysis of a small number of cells. It is unlikely that models of intracellular protein transport and trafficking will remain simple as knowledge of the processes grows and incorporates information on cell-to-cell variation. More sophisticated models that use nonlinear partial differential equations based on the geometry derived from imaging have been used to describe the intracellular dynamics of calcium and metabolites (Slepchenko et al., 2003). In each case, the main challenge in assessing the validity of quantitative predictions lies in careful analysis of the underlying assumptions, such as the use of Fickian diffusion to model the intracellular transport of proteins and small molecules.

Recent years have witnessed the discovery of a large number of highly organized, coherent, dissipative structure in cells, including waves of intracellular calcium and metabolites and protein concentration waves accompanying the division of bacterial cells (Kindzelskii and Petty, 2002; Schuster et al., 2002). While the general phenomenology of these structures is understood from the standpoint of nonlinear dynamics and physicochemical pattern formation, how these structures arise and how they are maintained and used by cells is a topic of intense research. Mathematical analysis of these processes requires significant extensions of the theory and computational methods for control of spatially distributed nonlinear systems.

In addition to studying and modeling intracellular structures, it is also important to model the possible mechanisms for their emergence from macromolecules. A number of experiments suggest that simple physicochemical principles can drive the emergence of cellular life (Szostak et al., 2001; Hanczyc et al., 2003; Chen et al., 2004). Indeed, experiments with nucleation of lipid vesicles by clays and the competition of protocells containing RNA polymerase suggest that the mechanisms of the formation of cells and intracellular compartments can be systematically studied in the test tube. Molecular simulations of these processes and population balance modeling of the evolution of primitive cellular compartments may provide the link between models at the molecular and cellular scales.

DISCOVERY OF CELLULAR NETWORKS AND THEIR FUNCTIONS

Since networks of interacting proteins control all cellular functions, understanding cellular functions requires quantitative analysis of the spatiotemporal dynamics of these networks in the cellular environment (Figure 4.3). Efforts to elucidate their dynamics can be subdivided into the analysis of network topology, dynamics, spatial organization, and function. Most of the progress recently has been in the area of deducing the network topology, using the classical techniques of cellular and molecular biology, large-scale molecular profiling experiments, or bioinformatics approaches (Brent and Finley, 1997; Chen and Xu, 2003; Ideker, 2004; Irish et al., 2004; Schulze and Mann, 2004; Xia et al., 2004; Yeger-Lotem et al., 2004). While there is significant room for the validation and perfection of each of these approaches, there is an urgent need to compare the networks predicted by distinct methods (Greenbaum et al., 2003). Because of the difficulties associated with generating high-quality data on cellular dynamics, much less work has been done in the analysis of network dynamics. For instance, some of the most interesting results associated with network dynamics have required construction of special experimental systems that include fluorescent reporters for a large number of bacterial genes coupled with high-resolution analysis of bacterial responses over a broad range of experimental conditions. This approach has led to the validation of the network motifs predicted on the basis of bioinformatics analysis and has identified the dynamic and functional roles of these motifs (Shen-Orr et al., 2002; Kalir and Alon, 2004; Zaslaver et al., 2004).

The simplest use of mathematical models for intracellular networks is to integrate data and test if they fit together. For example, mechanistic

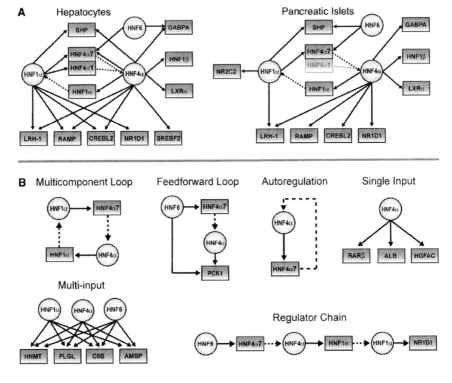

FIGURE 4.3 Transcriptional regulatory networks and motifs. (A) HNF1α, HNF6, and HNF4α are at the center of tissue-specific transcriptional regulatory networks. In these examples selected for illustration, regulatory proteins and their gene targets are represented as circles and boxes, respectively. Solid arrows indicate protein-DNA interactions, and genes encoding regulators are linked to their protein products by dashed lines. The HNF4α1 promoter is poorly expressed in pancreatic islets and is shaded to reflect this. The HNF4α7 promoter, also known as the P2 promoter, is the predominant promoter in pancreatic islets and was recently implicated as an important locus for human diabetes susceptibility. For, clarity, some gene promoters have been designated by the names of their protein products (e.g., HNF1α for *TCF1*, SHP for *NR0B2*, HNF4α7 for *HNF4A P2*, and HNF1β for *TCF2*). (B) Examples of regulatory network motifs in hepatocytes. For instance, in the multicomponent loop, HNF1α protein binds to the promoter of the HNF4α gene, and the HNF4α protein binds to the promoter of the HNF1α gene. These network motifs were uncovered by searching binding data with various algorithms; details on the algorithms used and a full list of motifs found are available. SOURCE: Odom et al., 2004.

models of the cell division cycle in fission yeast summarize the dynamic behavior of multiple mutants and make testable experimental predictions (Tyson et al., 2001). Most of the "realistic" models of intracellular networks, including the widely publicized models of the lysis-lysogeny switch in bacteriophage lambda and models of growth factor signaling, were severely overparameterized (Arkin et al., 1998; Schoeberl et al., 2002). On the one hand, the models have to be large to be able to predict the effects of genetic or biochemical perturbations of network components, and on the other, the models must have the lowest required number of parameters and processes to explain the observed phenomenology.

At this time, there are no standards for assessing the complexity of a model and whether it is indeed a minimal representation of data. Furthermore, current models are characterized by high levels of uncertainty, both structural and parametric. This situation is not surprising, given that even the most comprehensive models may be missing entire parts of a network, may neglect its spatial organization and temporal evolution, and may employ approximate functional forms for various cellular processes. While some tools for dealing with these issues can be borrowed from linear control systems (Csete and Doyle, 2002), new theoretical and computational approaches are required to analyze highly uncertain and nonlinear systems. These new approaches require solving problems in simulation, system identification, parameter estimation, and experimental design.

Recently, robustness has emerged as an important principle for model screening and validation (Barkai and Leibler, 1997; Stelling et al., 2004). In a nutshell, the relative plausibility of two models for the same process can be assessed by comparing the size of the parametric perturbations that can be tolerated by the models without qualitatively changing the predicted behavior. The rationale for using robustness as a screen is that evolution seems to favor the most robust mechanisms. For example, two models of the cell division cycle can be compared on the basis of the size of the regions of the parameter space that predict the limit cycle behavior (Morohashi et al., 2002). Analysis of robustness requires tools that can be used to compare models with different numbers of parameters and even different mathematical structures. The method of mathematically controlled comparisons is a very important development in this direction (Alves and Savageau, 2000). On the experimental side, the model-driven analysis of the robustness of cellular systems requires quantitative characterization of natural and induced parameter variations in cells (Houchmandzadeh et al., 2002; Jones et al., 2004). Quantitative and multivariable analysis of cell-to-cell variations is becoming possible thanks to advances in flow cytometry and live cell imaging (Irish et al., 2004).

The number of systems that have been characterized over the entire range from the biochemical description of the network to its dynamics and function is still very small. One of the best examples is the result of efforts by Ferrell and co-workers, who have analyzed the function of the mitogen-activated protein kinase (MAPK) network, a three-stage enzymatic cascade conserved from yeasts to humans. In an elegant sequence of papers, Ferrell et al. have shown that the multistage nature of the cascade enables it to act as a switch that is insensitive to small inputs at the top layer of the cascade and is fully activated when the threshold value for the input has been crossed (Huang and Ferrell, 1996; Ferrell, 1997; Ferrell and Machleder, 1998; Bagowski and Ferrell, 2001; Ferrell and Xiong, 2001; Xiong and Ferrell, 2003). This prediction, based on extensive computational analysis of the cascade model, has been validated through in vitro experiments with purified components of the network. The biochemical and modeling work set the stage for the analysis of MAPK dynamics in the frog oocyte maturation response. In this response system, it became apparent that the MAPK cascade is embedded in a positive feedback circuit that sharpens the threshold-detection capabilities of the circuit and mediates the irreversibility of the cell's maturation response to hormones. This insight was enabled by the large size of the frog oocyte, which made it possible to carry out single-cell biochemical assays of cellular responses, once again underscoring the importance of examining cellular responses at a single-cell level. Other examples of quantitative analysis of network dynamics at the single-cell level are now available (Irish et al., 2004; Jones et al., 2004; Lahav et al., 2004; Nelson et al., 2004; Raser and O'Shea, 2004).

The reductionist approach to cellular networks, which focuses on single modules such as the MAPK cascade or the EGF receptor pathway, has been reasonably successful. However, it must be realized that these modules do not operate in isolation and are affected by the large number of other processes occurring simultaneously. For example, genetic and biochemical evidence indicates that the MAPK cascade is coupled to essentially every other signal transduction pathway in cells. Quantitative understanding of cross talk in biochemical networks is necessary in order to probe these "cellular context" effects. While a modeling approach to the problem can start with simulations of coupled signaling or with genetic models, the eventual success of these models will depend on the availability of convenient experimental systems where pathway cross talk can be analyzed at the quantitative level.

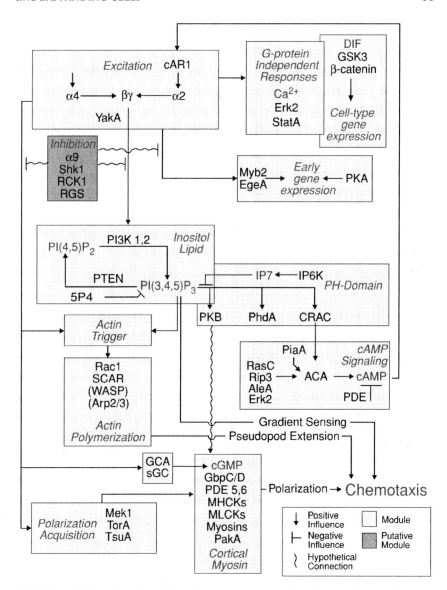

FIGURE 4.4 Modular view of the chemoattractant-induced signaling pathway in *Dictyostelium*. Except for those in parentheses, the proteins depicted in this pathway have been shown to be involved in chemotactic signaling through analysis of cells in which the genes have been deleted. SOURCE: Manahan et al., 2004.

FROM NETWORKS TO CELLULAR FUNCTIONS

After the analysis of gene and protein networks, the next goal in the quantitative understanding of cells is modeling integrated responses and functions, such as cell differentiation, migration, and the DNA damage response (Figure 4.4). Phenomenological models have been successful in correlating data and discovering qualitative trends in cellular responses. For example, simple biophysical and rheological models have been used to explain the biphasic dependence of cell migration speed on the adhesiveness of the substrate on which the cell is migrating (Lauffenburger and Horwitz, 1996). Simple birth-death processes have been used to model cell division and cell differentiation (Loeffler and Wichmann, 1980).

The earlier phenomenological models must be extended to integrate increasing amounts of information about each of these responses. In the case of cell migration, the intracellular rheology and motile behavior of cells is under the control of signal transduction and cytoskeletal networks that can now be monitored in real time and with increasing spatial resolution (Soll et al., 2000). Each of the biophysical and biochemical modules in cell migration, from signal transduction by integrins to the spatiotemporal dynamics of actin polymerization, is the subject of an extensive modeling effort (Grimm et al., 2003; Manahan et al., 2004). Notably, the modeling formalisms necessary to describe the integrated response have to be heterogeneous, in the sense that they differ in the amount of mechanistic detail incorporated in each specific model and in the mathematical structure of the model. For instance, a stochastic model of actin polymerization has to be coupled to deterministic models of the interactions between integrins and adhesive peptides. The main challenges for the development of quantitative models of cellular responses are related to the multiscale nature of each particular response and to the significant structural and parametric uncertainty of the current models. Integrated models of cell migration, currently being developed by the Cell Migration Consortium (http://www.cellmigration.org/), can be used as surrogates to test whether they can manifest the behavior predicted by earlier phenomenological models (Horwitz et al., 2002).

Perhaps the richest example of a cell-migration phenomenon that has been productively analyzed with the aid of quantitative models is bacterial chemotaxis (see Box 4.1). The ability of many bacteria to swim toward potential food sources or to evade noxious chemicals involves a complete cycle of signal detection and signal transduction and an elaborate response by cellular motility systems. To mount an effective chemotactic response, bacterial cells must not only detect the relevant chemical but also regulate behavior on the basis of spatial and temporal gradients in the signal. A different phenomenon, which also involves global changes in the regulation of bacterial genes in response to chemical signals, is quorum sensing (for a review, see Daniels et al., 2004). In this case, bacteria in a growing population both generate the chemical signal and respond to it. No one bacterium makes enough of the chemical to trigger the response system; however, once cell densities reach a threshold level (i.e., a "quorum" is achieved), the whole population of bacteria alters its regulatory state. This system, too, has attracted increasingly sophisticated mathematical modeling (Chopp et al., 2002; Ward et al., 2001, 2003). Some of this research is explicitly directed at exploring the potential of novel antibacterial drugs that would disrupt quorum sensing (Anguige et al., 2004); this concept is appealing since the regulatory change that many bacteria undergo when cell densities are high leads to increased expression of gene products that severely damage the tissues of an infected host.

Many instances of signal transduction and information processing in cells have been based on population-averaged data derived, for example, from Western blotting for the analysis of protein modification, as exemplified by Hoffmann et al. (2002) and Schoeberl et al. (2002). Currently, multicolor flow cytometry and high-throughput protein localization assays can monitor these processes at single-cell resolution and correlate the information obtained with responses measured at a single-cell level, such as the migratory tracks of individual cells or the differentiation responses of single cells. These data are enabling the systematic analysis of the heterogeneity of cellular responses (Krooshoop et al., 2003; Abraham et al., 2004; Irish et al., 2004). Statistical analysis of the resulting large sets of heterogeneous data—for instance, cell trajectories, protein modification, and localization—will become an important area of future research. In the end, models of cellular responses that can be used in biotechnology and medicine are likely to be reasonably simple correlations, similar in form to but more realistic than the phenomenological models of cellular responses developed over the past two decades.

The ultimate challenge in modeling cellular responses to signals is to track cause-and-effect relationships throughout the pathways leading from signal detection to cellular response. For most biological systems, this goal is remote. For relatively simple phenomena such as bacterial chemotaxis (see Box 4.1), many of cause-and-effect links, all the way from detection of chemical gradients to cellular locomotion, are now known. However, in systems of more typical complexity—particularly in multicellular organisms—the complexity on the response side of signal-response pathways poses immense challenges to modeling techniques. Progress appears likely to involve relatively detailed modeling of the upstream processes and, when possible, the final steps, such as cell motion. However, modeling of the linkage between the front and back ends of signal-response processes will often require heuristic methods that sim-

Box 4.1
Bacterial Chemotaxis

Bacterial chemotaxis in *Escherichia coli* (*E. coli*) is the best understood signal transduction system where one can go all the way from the molecular composition and subcellular organization of the biochemical network to the response of a single cell or population of cells (Berg, 2000). The current picture is a result of extensive genetic, biochemical, and biophysical analysis of this system over the past 50 years. The evolution of the quantitative understanding of bacterial chemotaxis shows unequivocally that helpful quantitative models are impossible without experimental innovations and are greatly enabled by the ease of genetic manipulation of the system.

The phenomenon was originally described by Adler at the macroscopic level as a directed flux of bacteria in gradients of nutrients such as aspartate (see Berg (2000) for a review). Berg and colleagues, who designed a tracking microscope that generated information about the migratory tracks of single cells, described the microscopic nature of bacterial chemotaxis. In isotropic environments, the path of *E. coli* is composed of straight runs punctuated by brief tumbles that can change the direction of migration. In a gradient of chemoattractant, the runs between the tumbling events are increased. At the population level, this change in the microscopic behavior of a single cell generates a directed flux of cells toward a source of a chemoattractant. Gradient sensing is mediated by ligand-receptor interaction at the cell surface that induces a sequence of biochemical reactions in the cytoplasm and culminates in the generation of the diffusible cytoplasmic molecule that binds to the motor powering the bacterial flagella, biases the sense of its rotation, and in this way changes the frequency of the tumbling events. The output of the circuit, defined as the frequency of the tumbling events, reflects the temporal derivative of receptor occupancy, which can be "measured" by bacteria with very high sensitivity and speed. The circuit is characterized by a very wide dynamic range, which is mediated by negative feedback loop in the signal transduction cascade. Today, we have detailed information about the genetics and biochemistry of the chemotaxis network in *E. coli*, we know the three-dimensional structures of several key proteins, we can monitor their interactions in vivo and in real time, and we can reconstitute parts of the network in vitro and measure the relevant thermodynamic and rate constants.

Bacterial chemotaxis has become a fruitful arena for modeling and computational analysis. Essentially every part of the circuit, from cell surface receptor to the flagellar motor, gave rise to mathematical models, ranging from structural descriptions of protein organization in the membrane (Bray and Duke, 2004), to the dynamics of signal transduction in the cytoplasm (Barkai and Leibler, 1997), to the kinetic theory of bacterial transport (Hillen and Othmer, 2000). The two most notable modeling efforts are the Berg and Purcell theory of ligand concentration measurement in gradient detection (Berg and Purcell, 1977) and the Barkai-Leibler model of robustness in the adaptation part of the circuit (Barkai and Leibler, 1997). Berg and Purcell used a model to test the hypothesis that bacteria can sense temporal gradients on the timescale of seconds. To test this hypothesis, they developed a very elegant stochastic biophysical theory of diffusion-limited ligand receptor binding. As a result of their analysis, they have concluded that bacteria can indeed measure concentrations and "take the temporal derivatives" of concentrations within a very short period of time. Their analysis was based on the assumption that receptors are distributed randomly over the surface of the cell. Subsequently their results were rederived by Szabo et al. (1982), using a much more transparent approach based on the homogenization theory and tested in direct Brownian dynamics simulations by Northrup (1988). The Berg-Purcell model of ligand-receptor binding in bacterial chemotaxis has become a classic in the theory of diffusion-limited reactions, continues to enjoy extensive citations, and has entered the textbooks in biophysics. At the same time, the electron microscopy images of E. coli produced in the early 1990s show that the main assumption of the theory, the random distribution of receptors, is not satisfied and that receptors are clustered in one region of the cell surface (Parkinson and Blair, 1993). Independent confirmations of this result and biochemical proof of receptor clustering gave rise to a new wave of models that attempt to explain their functional significance. At this time, it is established that receptor clustering is crucial for high sensitivity of the gradient sensing system. The dynamics and spatial organization of receptor clusters is now studied in models that are firmly grounded in the structural details of protein-protein interactions in bacterial chemotaxis (Bray and Duke, 2004).

While we are still a long way from having an integrated model of bacterial chemotaxis that would integrate all the structural, genetic, and biochemical evidence, analysis of this system over the past decades sets an excellent example of the most productive integration of experiments, multiscale biophysical modeling, and mathematical analysis (Erban and Othmer, 2005). One of the natural questions is whether the biophysical mechanisms of gradient detection in E.coli can be useful for understanding these processes in other cell types. A first step in this direction was taken in a recent computational model that compared the control strategies in E. coli and B. subtilis (Rao et al., 2004).

ply capture cause-and-effect correlations rather than providing quantitative models of actual pathways.

FROM CELLS TO TISSUES

Quantitative descriptions of cellular processes and functions form the basis for models at the tissue and organism level (Figure 4.5). Population-level modeling of cell migration is a good example of this approach (Maheshwari and Lauffenburger, 1998). Statistical analysis of the migratory tracks of single cells can be used to extract the probability density functions of cell velocities, turning frequencies, persistence time, and other variables. Such information about the properties of a single cell can be used to derive partial differential equations for the evolution of cell densities. The dynamics predicted by these equations can be quantitatively compared with the measurements of cellular fluxes (Hillen and Othmer, 2000). The same general framework, in which a microscopic description of particle motion is used to predict the evolution of particle ensembles, is encountered many times in the natural sciences—for example, in the derivation of the kinetic theory of gases or in the equations of fluid motion from the detailed description of molecular motion. The rules of cellular motion are much more complex than those governing the molecules in an ideal gas. However, cellular trajectories can be visualized with much greater ease than the trajectories of interacting molecules in gases or fluids (Othmer et al., 1988; Painter and Sherratt, 2003). An integrated program of this kind was implemented for bacterial and animal cells in the 1980s (Farrell et al., 1990). Today, similar analyses can be complemented with increasingly detailed information about the coupling between intracellular processes, such as signal transduction or cytoskeletal dynamics, and cellular responses, such as proliferation and migration. Multiscale models for the evolution of cell densities are being constructed to describe *E. coli* chemotaxis (Bren and Eisenbach, 2000; Erban and Othmer, 2005). Analysis of these models poses many challenging problems for multiscale theory and numerical analysis.

Modeling of tissue patterning is another example of analysis at the tissue level that is based on extensive studies of cellular processes. One of the mechanisms for generating cell diversity in embryogenesis is based on patterning of an epithelial layer, whereby a lattice of initially identical cells is presented with a spatial gradient of a ligand that binds to cell surface receptors and induces gene expression in target cells (Tabata and Takei, 2004). The level of cell surface receptor occupancy can be directly translated into the transcriptional response of the target cell. In this way, the spatial gradient of an extracellular ligand can be translated into a spatial pattern of gene expression in a layer of "naive" cells. Morphogen gra-

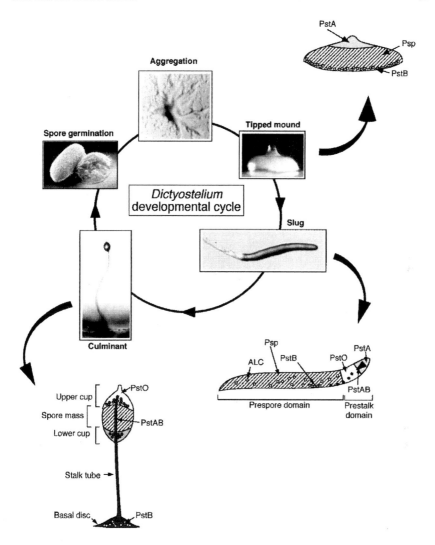

FIGURE 4.5 *Dictyostelium* development cycle and position of cell types within the multicellular differentiating organism. SOURCE: Kimmel and Firtel, 2004.

dients are established by the combination of localized secretion of ligands, their extracellular transport, binding to cell surface receptors, and intracellular trafficking processes. Computational models can be used to identify the relevant spatial and temporal scales in the generation of the morphogen gradients and to evaluate the relative feasibility of competing hypotheses (Lander et al., 2002; Kruse et al., 2004). These models can be based on earlier quantitative models of ligand-receptor dynamics in cells and on the current imaging experiments in developing tissues (Lauffenburger and Linderman, 1993).

Experimental models such as those for bacterial quorum sensing and slime mold aggregation enable biochemical and genetic analyses of the emergence of multicellular behavior in populations of seemingly identical cells (Taga and Bassler, 2003; Chisholm and Firtel, 2004). Detailed understanding of cell-to-cell communication is the key to developing integrative models of these processes (Nagano, 2000; Dockery and Keener, 2001). In all current biology textbooks, cell communication proceeds in a unidirectional way, where the signal is received by the cell and interpreted to direct cellular responses. In reality, cell-cell communication protocols are not unidirectional. Cells can both receive and respond to extracellular signals; hence they actively modify their environments. At the same time, cells are generating and responding to mixtures of signals. Modeling of such cell-cell communication protocols is complicated by the experimental difficulties associated with quantifying the dynamics and spatial regulation of cell communication signals. For example, mammalian cells can secrete a large number of soluble growth factors. The molecular identity of these signals and their potential roles in spatiotemporal information processing by cells is only beginning to be understood (Werb and Yan, 1998).

In addition to studying the processes through which single cells assemble into spatial patterns and tissues, it is important to study how tissues devolve to cells, as they do, for example, in the epithelial-mesenchymal transition (EMT), one of the critical steps in tumorigenesis. Genetic studies show that relatively small networks of genes can mediate EMT (Hahn and Weinberg, 2001). Translating this information into the integrative descriptions of epithelial dynamics poses an exciting and important problem for modeling.

DATA INTEGRATION

Rapid technical advances in genomics and proteomics have led to an extraordinary proliferation of data, which offers an unprecedented opportunity to understand how organisms function but poses significant challenges as well. Experimental design, hypothesis testing, and conceptual model building all require biologists to collect, evaluate, and integrate large amounts of information of many disparate kinds. There is a need to create user-friendly tools to assist researchers in designing and testing new hypotheses against the quickly growing, distributed knowledge base and to facilitate the development of experimentally verified models through the accumulation of validated hypotheses stored in databases designed from the ground up to support hypothesis testing.

Three kinds of conceptual and bioinformatics challenges appear in today's data-rich environment: (1) information retrieval and integration, (2) knowledge representation, and (3) hypothesis testing and model building. The first and second are closely related: How can we retrieve and express the many qualitatively different kinds of information available in databases and the published literature in a representation that informs experimentation? Developing common ontologies for biological objects and processes is essential for supporting the intercommunication of diverse databases (Schulze-Kremer, 1998; Ashburner et al., 2000) and for enabling the automated annotation and extraction of information from the published literature (Andrade et al., 1999; Fleischmann et al., 1999; Friedman et al., 2001; Stephens et al., 2001; Yakushiji et al., 2001). An ontology also provides the foundation for constructing higher-level representations of biological systems (Rzhetsky et al., 2000; Peleg et al., 2002).

The third challenge is to create and verify testable conceptual representations of the biological system. A conceptual framework for representing biological systems must accommodate the modularity and temporal evolution of biological networks, as well as handle their nonlinearity, plasticity, redundancy, and degeneracy. Conceptual models vary from the simple Boolean networks pioneered by Kaufmann (1969, 1993), Liang et al. (1998), and Akutsu et al. (2000a, 2000b) to Bayesian networks (Friedman et al., 2000; Hartemink et al., 2001; Pe'er et al., 2001), as well as highly concrete (McAdams and Arkin, 1998; Judd et al., 2000) and quantitative (Sveiczer et al., 2000) models. Incorporating disparate kinds of information about biological systems into a common conceptual framework remains a major stumbling block for validating ideas about real biological networks, and current efforts focus largely on just one or two categories of information (Rzhetsky et al., 2000; Hartemink et al., 2001; Wessels et al., 2001). There is a need to develop a hypothesis representation language that can assist in integrating experimental information at the logical level, as well as approaches to aggregating validated hypotheses into increasingly quantitative models.

Most currently available bioinformatics tools support the analytical tasks of the biologist. These tools are very useful and effective for certain specific tasks, such as identifying patterns, categorizing information, and simultaneously probing multiple data sources for similarities. Such tasks usually comprise the early steps of the discovery process. However, synthesis and evaluation of the information remain the task of the individual. Kuchinsky and his colleagues (2002) argue that this synthesis task can be broken down into steps: (1) keeping track of all the diverse pieces of information collected during the database searches and other retrieval activities, (2) organizing and using this information by formulating hypotheses and higher level explanations, and (3) sharing the information with colleagues and working collaboratively with colleagues to refine hypotheses. There is a need to develop a system to allow biologists to construct and verify formal language hypotheses, making use of event- and process-based description language.

Complex processes that exhibit nonlinear behavior, as biological systems do, are often more readily described by event-driven dynamics (Ho, 1989) than by differential equations. Furthermore, when biologists think about biological systems, they typically do so in terms of biological agents, events, and causal relationships between events. A symbolic discrete modeling approach, in which the discretizations are defined by events—that is, by any biological change for which there is experimental evidence of changes in the state of the system—is likely to be a useful approach.

BIOLOGICAL CONSIDERATIONS

The capacity of cells to differentiate is a hallmark of eukaryotic organisms. Differentiation is the acquisition of structurally and chemically different identities by cells over time. The capacity for self-differentiation, which transforms a single diploid cell (the zygote) into a complex, multicellular plant or animal comprising many different structural and functional tissues, is rooted in asymmetries within the initial cell and has its origins in the unequal intracellular distribution of small molecules, macromolecules, and organelles. Such inhomogeneities in cell structure can be triggered by external stimuli, such as fertilization (Green, 1993; Rossant and Tam, 2004; Swann et al., 2004) and light stimuli (Robinson et al., 1999), which set in motion a complex series of structural and compositional reorganizations that in turn generate compositionally different daughter cells, whose differences are further reinforced by differential gene expression (Kanka, 2003).

The rapid advances during the second half of the 20th century in the understanding of how DNA functions in heredity and expresses the information it encodes fostered a markedly genocentric view of eukaryotic development. This was further reinforced by a virtual flood of genome sequences and gene expression data that followed technological developments in nucleic acid sequencing and monitoring of gene expression patterns during the final decades of the century. Both embryonic development and cellular differentiation were viewed as under the control of genes, whose differential expression was orchestrated by a "developmental program." Indeed, the notion of a developmental program has dominated thinking about development for decades (Davidson et al., 1995; Marczynski and Shapiro, 1995; Chu et al., 1998; Roberts et al., 2000).

However, actual progress in understanding developmental processes has occurred by studying finer levels of detail rather than attempting to delineate an overarching developmental program. Mutations that affect development are often in genes that code for proteins that function in inter- and intracellular signaling and structure. Moreover, there are many molecular mechanisms by which cells affect the spatial patterning, including small molecules, such as the gaseous hormones nitric oxide (plants and animals) and ethylene (plants), and intermediate-sized molecules, such as the plant gibberellins and brassinosteroids and the animal endocrine and autocrine hormones.

Patterning mechanisms also include macromolecular mechanisms, such as the intracellular transport of proteins and RNA in both plants and animals. Examples include the translational gradients established in early *Drosophila* embryogenesis by the localization of *bicoid* and *oskar* mRNA molecules (Micklem et al., 2000) and the intercellular interactions that underlie the subsequent development of the abdominal segmentation pattern (Immergluck et al., 1990; Ingham and Arias, 1992; Courey and Huang, 1995). Analysis of processes such as these has led to articulation of the view that all of development can be explained by local interactions (Britten, 1998). Indeed, current developmental models are increasingly couched in terms of cellular fate determination by signaling between cells and by programmed cell death (Lam et al., 2001; Ribeiro et al., 2003; Lai and Orgogozo, 2004). Morphogenetic proteins can be secreted signaling proteins that alter the fates of cells through cell-surface-receptor-mediated pathways (Tabata and Takei, 2004). Receptors at the cell surface and intracellular signaling proteins, signaling cascades (such as MAPK cascades), and protein networks mediate the activation of genes in response to extracellular signals (Imler and Hoffmann, 2002; Muller and Bossinger, 2003; Schulz and Yutzey, 2004).

Epigenetic mechanisms are conceptualized as mechanisms that stably affect gene expression without altering gene structure. Initially, epigenetic mechanisms were equated with stable, even heritable, modifications in gene expression, commonly ones that suppress gene expression. Early descriptions of plant paramutation (Brink, 1960), transposable element inactivation (McClintock, 1965), and X-chromosome inactivation in mammals (Lyon, 1961) provided the foundation for what has become an active field of epigenetic research. Because differentiated plant cells can regenerate into whole plants and nuclei from differentiated animal cells can support development of enucleated eggs, DNA in most cells is not irreversibly altered during development and differentiation. However, the epigenetic modifications in gene expression that occur during development are highly regular, impose important differences between male and female gametes, and are not easily reversed experimentally

(Wrenzycki and Niemann, 2003; Tian, 2004) although they are altered regularly during gametogenesis and early embyronic development (Gehring et al., 2004; Santos and Dean, 2004).

There has been important progress in understanding the diverse mechanisms for epigenetic modifications of gene function. These mechanisms include DNA methylation, which compromises the digital purity of the Watson-Crick model of DNA by, in effect, putting "asterisks" on certain bits in the digital code, and highly dynamic chemical modifications of the histone proteins around which DNA wraps in the chromosomes of multicellular organisms. The complexity of these mechanisms and of the pathways that regulate their operation—together with the ubiquity of epigenetic effects during development—indicates that overly genecentric models of development are likely to fail.

FUTURE DIRECTIONS

The emphasis in cell biology is shifting from phenomenological descriptions to predictive models that are consistent with the largest possible amount of data. Data integration, reduction, and multiscale modeling approaches will take center stage in dealing with the diverse data sets emerging from molecular profiling and imaging experiments. Increasingly, biologists will use these approaches to identify the important species, interactions, and processes occurring within cells. Models of cellular processes must explicitly account for the significant parametric and structural uncertainties inevitable at the current level of knowledge and experimental resolution. In general, special attention should be paid to the proper selection and validation of the mathematical formalisms used to model any given process. Advances in dynamical systems theory will be required to deal with inevitable model heterogeneity, such as that required by the combination of stochastic descriptions of gene expression and deterministic descriptions of signal transduction events.

At this time, scientists are far from having an adequate quantitative description of cellular responses, even in well-studied systems such as bacterial chemotaxis. The field must make a concerted effort to improve the quantitative analysis of such model systems in order to achieve successes that can be used as templates for modeling and analysis in less well-established experimental systems. At the same time, the field must establish new experimental systems where integrative genetic, biochemical, and cell biological experiments are possible and that can support meaningful modeling efforts. It will be necessary to establish experimental systems that can serve as testing grounds for multiscale models that can be used to understand how cellular functions emerge from molecular-scale events and how cell population or tissue-level functions emerge from cellular-scale events.

Models must span as many scales as possible, from sequence-specific information on gene expression, to intracellular biochemistry, to cellular responses. Large-scale integrative approaches require the creation and funding of interacting groups of mathematicians, computer scientists, and biologists. While models should be based on detailed analysis of specific experimental systems—for example, particular cell types—model builders should strive to make the models generalizable to other systems. For example, it is important to analyze the evolution of cellular signaling systems in animals from worms to humans, as well as in plants. What mediates the increase in the number of signaling components in particular evolutionary linkages, and what are the systems-level consequences of this increase in complexity?

Quantitative experimental analyses of intracellular fluctuations and noise are critical for understanding cellular functions and the limits of applicability of conventional deterministic and continuum approaches. This type of analysis is also important for understanding the mechanisms and functional consequences of nongenetic individuality. Analysis of cell-to-cell variations is now possible owing to advances in imaging and single-cell molecular profiling experiments. On the purely theoretical and computational side, scientists must (1) classify dynamical systems according to the ways in which they can tolerate intracellular and extracellular noise and (2) understand the processes where noise can play a constructive role.

Thus, future research must emphasize the close connection between experiments, model validation, and data integration. While this level of integration is only possible by focusing on specific experimental systems, the field must make an effort to systematize methodological advances so that scientists do not have to start from scratch every time they analyze a new cellular system.

REFERENCES

Abraham, V.C., D.L. Taylor, J.R. Haskins. 2004. High content screening applied to large-scale cell biology. *Trends Biotechnol.* 22(1): 15-22.

Akutsu, T., S. Miyano, and S. Kuhara. 2000a. Algorithms for identifying Boolean networks and related biological networks based on matrix multiplication and fingerprint function. *J. Comput. Biol.* 7(3-4): 331-343.

Akutsu, T., S. Miyano, and S. Kuhara. 2000b. Inferring qualitative relations in genetic networks and metabolic pathways. *Bioinformatics* 16(8): 727-734.

Alves, R., and M.A. Savageau. 2000. Extending the method of mathematically controlled comparison to include numerical comparisons. *Bioinformatics* 16(9): 786-798.

Andrade, M.A., N.P. Brown, C. Leroy, S. Hoersch, A. de Daruvar, C. Reich, A. Franchini, J. Tamames, A. Valencia, C. Ouzounis, and C. Sander. 1999. Automated genome sequence analysis and annotation. *Bioinformatics* 15(5): 391-412.

Anguige, K., J.R. King, J.P. Ward, and P. Williams. 2004. Mathematical modelling of therapies targeted at bacterial quorum sensing. *Math. Biosci.* 192(1): 39-83.

Arkin, A., J. Ross, and H.H. McAdams. 1998. Stochastic kinetic analysis of developmental pathway bifurcation in phage lambda-infected Escherichia coli cells. *Genetics* 149(4): 1633-1648.

Ashburner, M., C.A. Ball, J.A. Blake, D. Botstein, H. Butler, J.M. Cherry, A.P. Davis, K. Dolinski, S.S. Dwight, J.T. Eppig, M.A. Harris, D.P. Hill, L. Issel-Tarver, A. Kasarskis, S. Lewis, J.C. Matese, J.E. Richardson, M. Ringwald, G.M. Rubin, and G. Sherlock. 2000. Gene ontology: Tool for the unification of biology. The Gene Ontology Consortium. *Nat. Genet.* 25(1): 25-29.

Bagowski, C.P., and J.E. Ferrell. 2001. Bistability in the JNK cascade. *Curr. Biol.* 11(15): 1176-1182.

Barkai, N., and S. Leibler. 1997. Robustness in simple biochemical networks. *Nature* 387(6636): 913-917.

Berg, H.C. 2000. Motile behavior of bacteria. *Physics Today* 53(1): 24-29.

Berg, H.C., and E.M. Purcell. 1977. Physics of chemoreception. *Biophys. J.* 20: 193-219.

Blagoev, B., I. Kratchmarova, S.-E. Ong, M. Nielsen, L.J. Foster, and M. Mann. 2003. A proteomics strategy to elucidate functional protein-protein interactions applied to EGF signaling. *Nat. Biotechnol.* 21(3): 315-318.

Bray, D., and T. Duke. 2004. Conformational spread: The propagation of allosteric states in large multiprotein complexes. *Annual Review of Biophysics and Biomolecular Structure* 33: 53-73.

Bren, A., and M. Eisenbach. 2000. How signals are heard during bacterial chemotaxis: Protein-protein interactions in sensory signal propagation. *J. Bacteriol.* 182(24): 6865-6873.

Brent, R., and R.L.J. Finley. 1997. Understanding gene and allele function with two-hybrid methods. *Annu. Rev. Genet.* 31: 663-704.

Brink, R.A. 1960. Paramutation and chromosome organization. *Q. Rev. Biol.* 35(2): 120-137.

Britten, R.J. 1998. Underlying assumptions of developmental models. *Proc. Natl. Acad. Sci. U.S.A.* 95(16): 9372-9377.

Chen, I.A., R.W. Roberts, J.W. Szostak. 2004. The emergence of competition between model protocells. *Science* 305(5689): 1474-1476.

Chen, Y., and D. Xu. 2003. Computational analyses of high-throughput protein-protein interaction data. *Curr. Protein Pept. Sci.* 4(3): 159-181.

Chisholm, R.L., and R.A. Firtel. 2004. Insights into morphogenesis from a simple developmental system. *Nat. Rev. Mol. Cell. Biol.* 5(7): 531-541.

Chopp, D.L., M.J. Kirisits, B. Moran, and M.R. Parsek. 2002. A mathematical model of quorum sensing in a growing bacterial biofilm. *J. Ind. Microbiol. Biot.* 29(6): 339-346.

Chu, S., J. DeRisi, M. Eisen, J. Mulholland, D. Botstein, P.O. Brown, and I. Herskowitz. 1998. The transcriptional program of sporulation in budding yeast. *Science* 282(5389): 699-705.

Courey, A.J., and J.D. Huang. 1995. The establishment and interpretation of transcription factor gradients in the Drosophila embryo. *BBA-Gene Struct. Expr.* 1261(1): 1-18.

Csete, M.E., and J.C. Doyle. 2002. Reverse engineering of biological complexity. *Science* 295(5560): 1664-1669.

Daniels, R., J. Vanderleyden, and J. Michiels. 2004. Quorum sensing and swarming migration in bacteria. *FEMS Microbiol. Rev.* 28(3): 261-289.

Davidson, E.H., K.J. Peterson, and R.A. Cameron. 1995. Origin of bilaterian body plans: Evolution of developmental regulatory mechanisms. *Science* 270(5240): 1319-1325.

Dockery, J.D., and J.P. Keener. 2001. A mathematical model for quorum sensing in Pseudomonas aeruginosa. *Bull. Math. Biol.* 63(1): 95-116.

Erban, R., and H.G. Othmer. 2005. From signal transduction to spatial pattern formation in *E. coli*: A paradigm for multi-scale modeling in biology. *Multiscale Model. Simul.* 3(2): 362-394.

Farrell, B.E., R.P. Daniele, D.A. Lauffenburger. 1990. Quantitative relationships between single-cell and cell-population model parameters for chemosensory migration responses of alveolar macrophages to C5a. *Cell Motil. Cytoskeleton* 16(4): 279-293.

Ferrell, J. 1997. How responses get more switch-like as you move down a protein kinase cascade. *Trends Biochem. Sci.* 22(8): 288-289.

Ferrell, J., and W. Xiong. 2001. Bistability in cell signaling: How to make continuous processes discontinuous, and reversible processes irreversible. *Chaos* 11(1): 221-236.

Ferrell, J.J., and E. Machleder. 1998. The biochemical basis of an all-or-none cell fate switch in Xenopus oocytes. *Science* 280(5385): 895-898.

Fleischmann, W., S. Moller, A. Gateau, and R. Apweiler. 1999. A novel method for automatic functional annotation of proteins. *Bioinformatics* 15(3): 228-233.

Friedman, C., P. Kra, H. Yu, M. Krauthammer, and A. Rzhetsky. 2001. GENIES: A natural-language processing system for the extraction of molecular pathways from journal articles. *Bioinformatics* 17(Suppl 1): S74-S82.

Friedman, N., M. Linial, I. Nachman, and D. Pe'er. 2000. Using Bayesian networks to analyze expression data. *J. Comput. Biol.* 7(3-4): 601-620.

Gehring, M., Y. Choi, and R.L. Fischer. 2004. Imprinting and seed development. *Plant Cell* 16: S203-S213.

Green, D.P. 1993. Mammalian fertilization as a biological machine: A working model for adhesion and fusion of sperm and oocyte. *Hum. Reprod.* 8(1): 91-96.

Greenbaum, D., C. Colangelo, K. Williams, and M. Gerstein. 2003. Comparing protein abundance and mRNA expression levels on a genomic scale. *Genome Biol.* 4(9): 117.

Grimm, H.P., A.B. Verkhovsky, A. Mogilner, and J.J. Meister. 2003. Analysis of actin dynamics at the leading edge of crawling cells: Implications for the shape of keratocyte lamellipodia. *Eur. Biophys. J.* 32(6): 563-577.

Hahn, W.C., and R.A. Weinberg. 2001. Rules for making human tumor cells. *N. Engl. J. Med.* 347(20): 1593-1603.

Hanczyc, M.M., S.M. Fujikawa, and J.W. Szostak. 2003. Experimental models of primitive cellular compartments: Encapsulation, growth, and division. *Science* 302(5645): 618-622.

Hartemink, A.J., D.K. Gifford, T.S. Jaakkola, and R.A. Young. 2001. Using graphical models and genomic expression data to statistically validate models of genetic regulatory networks. Pp. 422-433 in *Biocomputing 2001: Proceedings of the Pacific Symposium*. Singapore: World Scientific.

Hillen, T., and H.G. Othmer. 2000. Chemotaxis equations from the diffusion limit of transport equations. *SIAM J. Appl. Math.* 62: 1222-1250.

Hirschberg, K., C.M. Miller, J. Ellenberg, J.F. Presley, E.D. Siggia, R.D. Phair, and J. Lippincott-Schwartz. 1998. Kinetic analysis of secretory protein traffic and characterization of golgi to plasma membrane transport intermediates in living cells. *J. Cell Biol.* 143(6): 1485-1503.

Ho, Y.C. 1989. Special issue on discrete event dynamical systems: Editorial. *Proc. IEEE* 77(1): 24-38.

Hoffmann, A., A. Levchenko, M.L. Scott, and D. Baltimore. 2002. The IκB-NF-κB signaling module: Temporal control and selective gene activation. *Science* 298(5596): 1241-1245.

Horwitz, A.R., N. Watson, and J.T. Parsons. 2002. Breaking barriers through collaboration: The example of the Cell Migration Consortium. *Genome Biol.* 3(11): comment 2011.

Houchmandzadeh, B., E. Wieschaus, and S. Leibler. 2002. Establishment of developmental precision and proportions in the early Drosophila embryo. *Nature* 415(6873): 798-802.

Huang, C., and J.J. Ferrell. 1996. Ultrasensitivity in the mitogen-activated protein kinase cascade. *Proc. Natl. Acad. Sci. U.S.A.* 93(19): 10078-10083.

Ideker, T. 2004. A systems approach to discovering signaling and regulatory pathways—or, how to digest large interaction networks into relevant pieces. *Adv. Exp. Med. Biol.* 547: 21-30.

Imler, J.L., and J.A. Hoffmann. 2002. Toll receptors in Drosophila: A family of molecules regulating development and immunity. *Curr. Top. Microbiol.* 270: 63-79.

Immergluck, K., P.A. Lawrence, and M. Bienz. 1990. Induction across germ layers in Drosophila mediated by a genetic cascade. *Cell* 62(2): 261-268.

Ingham, P.W., and A.M. Arias. 1992. Boundaries and fields in early embryos. *Cell* 68(2): 221-235.

Irish, J.M., R. Hovland, P.O. Krutzik, O.D. Perez, O. Bruserud, B.T. Gjertsen, and G.P. Nolan. 2004. Single cell profiling of potentiated phospho-protein networks in cancer cells. *Cell* 118(2): 217-228.

Jones, J.T., J.W. Myers, J.E. Ferrell, and T. Meyer. 2004. Probing the precision of the mitotic clock with a live-cell fluorescent biosensor. *Nat. Biotechnol.* 22(3): 306-312.

Judd, E.M., M.T. Laub, H.H. McAdams. 2000. Toggles and oscillators: New genetic circuit designs. *Bioessays* 22(6): 507-509.

Kalir, S., and U. Alon. 2004. Using a quantitative blueprint to reprogram the dynamics of the flagella gene network. *Cell* 117(6): 713-720.

Kanka, J. 2003. Gene expression and chromatin structure in the pre-implantation embryo. *Theriogenology* 59(1): 3-19.

Kauffman, S.A. 1993. *The Origins of Order: Self-organization and Selection in Evolution.* New York, N.Y.: Oxford University Press.

Kaufmann, S.A. 1969. Metabolic stability and epigenesis in randomly connected nets. *J. Theor. Biol.* 22: 437-467.

Kimmel, A.R., and R.A. Firtel. 2004. Breaking symmetries: Regulation of *Dictyostelium* development through chemoattractant and morphogen signal-response. *Curr. Opin. Genet. Dev.* 14(5): 540-549.

Kindzelskii, A.L., and H.R. Petty. 2002. Apparent role of traveling metabolic waves in oxidant release by living neutrophils. *Proc. Natl. Acad. Sci. U.S.A.* 99(14): 9207-9212.

Krooshoop, D.J., R. Torensma, G.J. van den Bosch, J.M. Nelissen, C.G. Figdor, R.A. Raymakers, and J.B. Boezeman. 2003. An automated multi well cell track system to study leukocyte migration. *J. Immunol. Methods* 280(1-2): 89-102.

Kruse, K., P. Pantazis, T. Bollenbach, F. Jülicher, and M. González-Gaitán. 2004. Dpp gradient formation by dynamin-dependent endocytosis: Receptor trafficking and the diffusion model. *Development* 131(19): 4843-4856.

Kuchinsky, A., K. Graham, D. Moh, A. Adler, K. Babaria, and M.L. Creech. 2002. Biological storytelling: A software tool for biological information organization based upon narrative structure. *ACM SIGGROUP Bulletin* 23(2): 4-5.

Lahav, G., N. Rosenfeld, A. Sigal, N. Geva-Zatorsky, A.J. Levine, M.B. Elowitz, and U. Alon. 2004. Dynamics of the p53-Mdm2 feedback loop in individual cells. *Nat. Genet.* 36(2): 147-150.

Lai, E.C., and V. Orgogozo. 2004. A hidden program in Drosophila peripheral neurogenesis revealed: Fundamental principles underlying sensory organ diversity. *Dev. Biol.* 269(1): 1-17.

Lam, E., N. Kato, and M. Lawton. 2001. Programmed cell death, mitochondria and the plant hypersensitive response. *Nature* 411(6839): 848-853.

Lander, A.D., W. Nie, and F.Y.-M. Wan. 2002. Do morphogen gradients arise by diffusion? *Dev. Cell* 2(6): 785-796.

Lauffenburger, D.A., and A.F. Horwitz. 1996. Cell migration: A physically integrated molecular process. *Cell* 84(3): 359-369.

Lauffenburger, D.A., and J.J. Linderman. 1993. *Receptors: Models for Binding, Trafficking, and Signaling.* New York, N.Y.: Oxford University Press.

Liang, S., S. Fuhrman, and R. Somogyi. 1998. REVEAL, a general reverse engineering algorithm for inference of genetic network architectures. Pp. 18-29 in *Biocomputing '98: Proceedings of the Pacific Symposium.* Singapore: World Scientific.

Loeffler, M., and H.E. Wichmann. 1980. A comprehensive mathematical model of stem cell proliferation which reproduces most of the published experimental results. *Cell Tiss. Kinet.* 13: 543-561.

Lyon, M. 1961. Gene action in the X-chromosome of the mouse. *Nature* 190: 372-373.

Maheshwari, G., and D.A. Lauffenburger. 1998. Deconstructing (and reconstructing) cell migration. *Microsc. Res. Tech.* 43(5): 358-368.

Manahan, C.L., P.A. Iglesias, Y. Long, and P.N. Devreotes. 2004. Chemoattractant signaling in dictyostelium discoideum. *Annu. Rev. Cell Dev. Biol.* 20: 223-253.

Marczynski, G.T., and L. Shapiro. 1995. The control of asymmetric gene-expression during caulobacter cell-differentiation. *Arch. Microbiol.* 163(5): 313-321.

McAdams, H.H., and A. Arkin. 1998. Simulation of prokaryotic genetic circuits. *Annu. Rev. Biophys. Biomol. Struct.* 27: 199-224.

McClintock, B. 1965. The control of gene action in maize. Pp. 162-184 in *Brookhaven Symposium in Biology.* Upton, N.Y.: Brookhaven National Laboratory.

Micklem, D.R., J. Adams, S. Grunert, and D. St. Johnston. 2000. Distinct roles of two conserved Staufen domains in oskar mRNA localization and translation. *EMBO J.* 19(6): 1366-1377.

Morohashi, M., A.E. Winn, M.T. Borisuk, H. Bolouri, J. Doyle, and H. Kitano. 2002. Robustness as a measure of plausibility in models of biochemical networks. *J. Theor. Biol.* 216(1): 19-30.

Muller, H.A.J., and O. Bossinger. 2003. Molecular networks controlling epithelial cell polarity in development. *Mech. Dev.* 120(11): 1231-1256.

Nagano, S. 2000. Modeling the model organism Dictyostelium discoideum. *Dev. Growth Differ.* 42(6): 541-550.

Nelson, D.E., A.E. Ihekwaba, M. Elliott, J.R. Johnson, C.A. Gibney, B.E. Foreman, G. Nelson, V. See, C.A. Horton, D.G. Spiller, S.W. Edwards, H.P. McDowell, J.F. Unitt, E. Sullivan, R. Grimley, N. Benson, D. Broomhead, D.B. Kell, and M.R.H. White. 2004. Oscillations in NF-κB signaling control the dynamics of gene expression. *Science* 306(5696): 704-708.

Northrup, S.H. 1988. Diffusion-controlled ligand binding to multiple competing cell-bound receptors. *J. Phys. Chem.* 92: 5847-5850.

Odom, D.T., N. Zizlsperger, D.B. Gordon, G.W. Bell, N.J. Rinaldi, H.L. Murray, T.L. Volkert, J. Schreiber, P.A. Rolfe, D.K. Gifford, E. Fraenkel, G.I. Bell, and R.A. Young. 2004. Control of pancreas and liver gene expression by HNF transcription factors. *Science* 303(5662): 1378-1381.

Othmer, H.G., S.R. Dunbar, and W. Alt. 1988. Models of dispersal in biological systems. *J. Math. Biol.* 26(3): 263-298.

Painter, K.J., and J.A. Sherratt. 2003. Modelling the movement of interacting cell populations. *J. Theor. Biol.* 225(3): 327-339.

Parkinson, J.S., and D.F. Blair. 1993. Does *E. coli* have a nose? *Science* 259: 1701-1702.

Pe'er, D., A. Regev, G. Elidan, and N. Friedman. 2001. Inferring subnetworks from perturbed expression profiles. *Bioinformatics* 17(Suppl. 1): 215-224.

Peleg, M., I. Yeh, and R.B. Altman. 2002. Modelling biological processes using workflow and Petri Net models. *Bioinformatics* 18(6): 825-837.

Rao, C.V., J.R. Kirby, and A.P. Arkin. 2004. Design in bacterial chemotaxis: A comparative study in Escherichia coli and Bacillus subtilis. *PLOS Biology* 2(2): 239-252.

Raser, J.M., and E.K. O'Shea. 2004. Control of stochasticity in eukaryotic gene expression. *Science* 304(5678): 1811-1814.

Ribeiro, C., V. Petit, and M. Affolter. 2003. Signaling systems, guided cell migration, and organogenesis: Insights from genetic studies in Drosophila. *Dev. Biol.* 260(1): 1-8.

Roberts, C.J., B. Nelson, M.J. Marton, R. Stoughton, M.R. Meyer, H.A. Bennett, Y.D. He, H. Dai, W.L. Walker, T.R. Hughes, M. Tyers, C. Boone, and S.H. Friend. 2000. Signaling and circuitry of multiple MAPK pathways revealed by a matrix of global gene expression profiles. *Science* 287(5454): 873-880.

Rossant, J., and P.P.L. Tam. 2004. Emerging asymmetry and embryonic patterning in early mouse development. *Dev. Cell* 7(2): 155-164.

Rzhetsky, A., T. Koike, S. Kalachikov, S.M. Gomez, M. Krauthammer, S.H. Kaplan, P. Kra, J.J. Russo, and C. Friedman. 2000. A knowledge model for analysis and simulation of regulatory networks. *Bioinformatics* 16(12): 1120-1128.

Santos, F., and W. Dean. 2004. Epigenetic reprogramming during early development in mammals. *Reproduction* 127(6): 643-651.

Schoeberl, B., C. Eichler-Jonsson, E.D. Gilles, and G. Müller. 2002. Computational modeling of the dynamics of the MAP kinase cascade activated by surface and internalized EGF receptors. *Nat. Biotechnol.* 20(4): 370-375.

Schulz, R.A., and K.E. Yutzey. 2004. Calcineurin signaling and NFAT activation in cardiovascular and skeletal muscle development. *Dev. Biol.* 266(1): 1-16.

Schulze, W.X., and M. Mann. 2004. A novel proteomic screen for peptide-protein interactions. *J. Biol. Chem.* 279(11): 10756-10764.

Schulze-Kremer, S. 1998. Ontologies for molecular biology. Pp. 693-704 in *Biocomputing '98: Proceedings of the Pacific Symposium.* Singapore: World Scientific.

Schuster, S., M. Marhl, and T. Hofer. 2002. Modelling of simple and complex calcium oscillations: From single-cell responses to intercellular signalling. *Eur. J. Biochem.* 269(5): 1333-1355.

Shen-Orr, S.S., R. Milo, S. Mangan, and U. Alon. 2002. Network motifs in the transcriptional regulation network of Escherichia coli. *Nat. Genet.* 31(1): 64-68.

Slepchenko, B.M., J.C. Schaff, I. Macara, and L.M. Loew. 2003. Quantitative cell biology with the Virtual Cell. *Trends Cell Biol.* 13(11): 570-576.

Smith, A.E., B.M. Slepchenko, J.C. Schaff, L.M. Loew, and I.G. Macara. 2002. Systems analysis of Ran transport. *Science* 295(5554): 488-491.

Soll, D.R., E. Voss, O. Johnson, and D. Wessels. 2000. Three-dimensional reconstruction and motion analysis of living, crawling cells. *Scanning* 22(4): 249-257.

Sorkin, A., M. McClure, F. Huang, and R. Carter. 2000. Interaction of EGF receptor and grb2 in living cells visualized by fluorescence resonance energy transfer (FRET) microscopy. *Curr. Biol.* 10(21): 1395-1398.

Stelling, J., U. Sauer, Z. Szallasi, F.J. Doyle III, and J. Doyle. 2004. Robustness of cellular functions. *Cell* 118(6): 675-685.

Stephens, M., M. Palakal, S. Mukhopadhyay, R. Raje, and J. Mostafa. 2001. Detecting gene relations from Medline abstracts. Pp. 483-495 in *Biocomputing 2001: Proceedings of the Pacific Symposium.* Singapore: World Scientific.

Sveiczer, A., A. Csikasz-Nagy, B. Gyorffy, J.J. Tyson, and B. Novak. 2000. Modeling the fission yeast cell cycle: Quantized cycle times in wee1-cdc25Delta mutant cells. *Proc. Natl. Acad. Sci. U.S.A.* 97(14): 7865-7870.

Swann, K., M.G. Larman, C.M. Saunders, and F.A. Lai. 2004. The cytosolic sperm factor that triggers Ca2+ oscillations and egg activation in mammals is a novel phospholipase C: PLCzeta. *Reproduction* 127(4): 431-439.

Swedlow, J.R., I. Goldberg, E. Brauner, and P.K. Sorger. 2003. Informatics and quantitative analysis in biological imaging. *Science* 300(5616): 100-102.

Szabo, A., D. Shoup, S.H. Northrup, and J.A. McCammon. 1982. Stochastically gated diffusion-influenced reactions. *J. Chem. Phys.* 77: 4484-4493.

Szostak, J.W., D.P. Bartel, and L. Luisi. 2001. Synthesizing life. *Nature* 409(6818): 387-390.

Tabata, T., and Y. Takei. 2004. Morphogens, their identification and regulation. *Development* 131(4): 703-712.

Taga, M.E., and B.L. Bassler. 2003. Chemical communication among bacteria. *Proc. Natl. Acad. Sci. U.S.A.* 100(2): 14549-14554.

Tian, X.C. 2004. Reprogramming of epigenetic inheritance by somatic cell nuclear transfer. *Reprod. Biomed. Online* 8(5): 501-508.

Tyson, J.J., K. Chen, and B. Novak. 2001. Network dynamics and cell physiology. *Nat. Rev. Mol. Cell Biol.* 2(12): 908-916.

Ward, J.P., J.R. King, A.J. Koerber, J.M. Croft, R.E. Sockett, and P. Williams. 2003. Early development and quorum sensing in bacterial biofilms. *J. Math. Biol.* 47(1): 23-55.

Ward, J.P., J.R. King, A.J. Koerber, P. Williams, J.M. Croft, and R.E. Sockett. 2001. Mathematical modelling of quorum sensing in bacteria. *IMA J. Math. Appl. Med. Biol.* 18(3): 263-292.

Werb, Z., and Y.B. Yan. 1998. Cell biology: A cellular striptease act. *Science* 282(5392): 1279-1280.

Wessels, L.F., E.P. van Someren, and M.J. Reinders. 2001. A comparison of genetic network models. Pp. 508-519 in *Biocomputing 2001: Proceedings of the Pacific Symposium.* Singapore: World Scientific.

Wiley, H.S. 2003. Trafficking of the ErbB receptors and its influence on signaling. *Exp. Cell Res.* 284(1): 78-88.

Wiley, H.S., and D.D. Cunningham. 1981. A steady-state model for analyzing the cellular-binding, internalization, and degradation of polypeptide ligands. *Cell* 25(2): 433-440.

Wiley, H.S., J.J. Herbst, B.J. Walsh, D.A. Lauffenburger, M.G. Rosenfeld, and G.N. Gill. 1991. The role of tyrosine kinase activity in endocytosis, compartmentation, and down-regulation of the epidermal growth factor receptor. *J. Biol. Chem.* 266(17): 11083-11094.

Wiley, H.S., S.Y. Shvartsman, and D.A. Lauffenburger. 2003. Computational modeling of the EGF-receptor system: A paradigm for systems biology. *Trends Cell Biol.* 13(1): 43-50.

Wrenzycki, C., and H. Niemann. 2003. Epigenetic reprogramming in early embryonic development: Effects of in-vitro production and somatic nuclear transfer. *Reprod. Biomed. Online* 7(6): 649-656.

Xia, Y., H. Yu, R. Jansen, M. Seringhaus, S. Baxter, D. Greenbaum, H. Zhao, and M. Gerstein. 2004. Analyzing cellular biochemistry in terms of molecular networks. *Annu. Rev. Biochem.* 73: 1051-1087.

Xiong, W., and J.E.J. Ferrell. 2003. A positive-feedback-based bistable 'memory module' that governs a cell fate decision. *Nature* 426(6925): 460-465.

Yakushiji, A., Y. Tateisi, Y. Miyao, and J. Tsujii. 2001. Event extraction from biomedical papers using a full parser. Pp. 408-419 in *Biocomputing 2001: Proceedings of the Pacific Symposium.* Singapore: World Scientific.

Yeger-Lotem, E., S. Sattath, N. Kashtan, S. Itzkovitz, R. Milo, R.Y. Pinter, U. Alon, and H. Margalit. 2004. Network motifs in integrated cellular networks of transcription-regulation and protein-protein interaction. *Proc. Natl. Acad. Sci. U.S.A.* 101(16): 5934-5939.

Young, D.W., S.K. Zaidi, P.S. Furcinitti, A. Javed, A.J. van Wijnen, J.L. Stein, J.B. Lian, and G.S. Stein. 2004. Quantitative signature for architectural organization of regulatory factors using intranuclear informatics. *J. Cell Sci.* 117(21): 4889-4896.

Zaslaver, A., A.E. Mayo, R. Rosenberg, P. Bashkin, H. Sberro, M. Tsalyuk, M.G. Surette, and U. Alon. 2004. Just-in-time transcription program in metabolic pathways. *Nat. Genet.* 36(5): 486-491.

Zilberman, D., X.F. Cao, L.K. Johansen, Z. Xie, J.C. Carrington, and S.E. Jacobsen. 2004. Role of Arabidopsis ARGONAUTE4 in RNA-directed DNA methylation triggered by inverted repeats. *Curr. Biol.* 14(13): 1214-1220.

5

Understanding Organisms

The step in hierarchical scale from cells to organisms is huge. This chapter largely discusses tissues, organs, and organ systems in multicellular organisms made up of immense numbers of highly differentiated cells. In some instances, as in the discussion of locomotion, the focus is on the integrated properties of the whole organism. While much mathematical analysis directed at understanding organisms involves systems that are far removed from the cellular level, there are also levels of biological organization intermediate between cells and organisms. Biofilms formed by bacteria are one such intermediate level that has received modeling attention (Morgenroth et al., 2004), as are still more complex cellular aggregates such as the slug phase of the slime mold *Dictyostelium discoideum* (Umeda and Inouye, 2004) and the aggegration phase of myxobacteria (Igoshin and Oster, 2004). Analysis of systems of this type—poised as they are between cells and organisms on the scale of biological organization—offers a promising path toward improving mathematical approaches to multicellular processes. However, the committee draws its main examples in this chapter from more traditional areas of mathematical modeling, such as physiological processes.

In recent years the importance of mathematical models in the study of physiological processes has become widely accepted. There are many instances of how experimentalists and theoreticians, working together, have made discoveries that would be difficult, if not impossible, for each working independently. One such discovery involves the phenomenon of electrical excitability and the propagation of action potentials in cardiac and neural tissue. How oscillations in the cell cycle lead to regular cell divisions; how intercellular calcium waves coordinate cellular responses over large areas; how tumors grow and respond to chemotherapy; and how the HIV virus is produced and cleared within cells: all are areas where mathematical models have played an important role.

The need for mathematical models has never been greater. Much of the biological investigation of the past can be described as a compilation and categorization of the list of parts, whether as the delineation of genomic sequences, genes, proteins, or species. The past decade has seen an explosion in probing genetic or cellular defects that alter properties and behaviors at the tissue or organ level, thereby identifying the root basis for many diseases. As examples, we know the mutation to a chloride ion channel that results in cystic fibrosis and the mutations to potassium channels that lead to long QT syndrome (an abnormality of the heart's electrical system). There have also been many striking advances in imaging and measurement of function, some due to mathematical advances that provide insight into the level and extent of functional degradation or guide clinical intervention. For example, the ability to interpret electrocardiograms has led to spectacular advances in the reliability of implantable pacemakers and defibrillators (Kenknight et al., 1996). Missing is the ability to integrate how the various components of organs work together to achieve dynamic function, and how change of specific components or combinations thereof impact function. Thus, the challenge of systems physiology is to provide an understanding of how the interactions of biological entities across spatial and temporal scales lead to observable behavior and function.

Two important organizing principles need emphasis. First, an integrated understanding of systems requires mathematics and the development of theory, supplemented by simulations. One of the important lessons of the past is that there are behaviors and phenomena that are the consequences of interactions of several or many individual components that cannot occur with the components uncoupled, and the principles governing these emergent behaviors require theory for their full explanation. Secondly, theory cannot be relevant if it is not driven and inspired by experimental data. The committee illustrates these with some examples where systems physiology has great promise.

CARDIAC PHYSIOLOGY

Failure of the cardiac system remains the leading cause of death in the Western world. The cardiac cycle consists of two primary events: (1) a contractile, or mechanical, event, controlled by (2) an electrical event, the cardiac action potential. Failure of either of these can lead to death. Either cardiomyopathies, in which the cardiac muscle does not provide enough force or blood volume (decreased cardiac output), or a disruption of the electrical signal in an otherwise mechanically adequate heart may be the primary dysfunction. Of course, these disruptions are rarely completely independent, as diseased and damaged tissue often results in an increased likelihood of electrical malfunction.

One challenge presented by the cardiac system is to understand the physiological mechanisms underlying the electrical signal, so as to understand the mechanisms of the variety of arrhythmias and to learn how to control or prevent these arrhythmias. A substantial amount of ongoing research is aimed at understanding the dynamics of cardiac cells using mathematical and computational models. There is a long history to this direction of investigation, which has its origins with the Hodgkin-Huxley equations. The Hodgkin-Huxley theory was extended to cardiac cells by Noble, Beeler-Reuter, and others. More recently, detailed cellular ionic models have been developed by, for instance, Luo and Rudy (1994), Jafri et al. (1998), and Puglisi and Bers (2001).

In spite of the remarkable success of these models, they all fall short of providing an understanding of many important arrhythmias. This shortcoming is illustrated by the history of antiarrhythmic drugs. Many of the so-called antiarrhythmic drugs are known to be ion channel blockers. When they were first discovered, it was thought that arrhythmias were caused by overactive ion channels and if these were blocked, then the arrhythmias could be prevented. Indeed, tests on single cells and small patches of tissue verified this conjecture. However, when drugs were tested in the CAST and SWORD clinical trials (CAST Investigators, 1989; Waldo et al., 1996), it was discovered that many of these drugs were actually proarrhythmic. The fundamental difficulty was that an understanding of how single cells or small patches of tissue behave or respond to drugs does not answer the question of how the entire spatiotemporal system will behave. (While some arrhythmias are the result of cellular automaticity and ectopic foci, which occur when a cell or small collection of cells oscillates without external stimulus and thereby takes over as the pacemaker of the heart, the most significant life-threatening arrhythmias are maintained because they are spatiotemporal patterns and cannot occur in single cells or small patches of tissue.) In the case of the CAST and SWORD studies, the lack of a suitable spatiotemporal model led people astray. They relied instead on their best guess, but their best guess was wrong. It is now recognized that almost all drugs that were previously classified as antiarrhythmic are actually proarrhythmic. It is also now recognized that the response of a single cell to ion-channel blockers does not adequately predict the response at the spatiotemporal level.

Thus, the challenge is to develop mechanistic, functionally integrated, multiscale mathematical models of the heart from molecular to cellular and whole-organ scales, which would lead to a deeper understanding of the excitation and contraction of the heart. Research needs to move from understanding atrial and ventricular electrophysiology based on models of the biophysics of single-ion channels to predicting the electrocardiogram recorded at the body surface. (Of course we will ultimately want to use this understanding in an inverse way: interpreting electrocardiogram signals at the surface as indicators of the functioning of cardiac subsystems.) An overarching theme will be how mathematical models can help elucidate mechanisms, improve diagnoses, and identify therapeutic targets for cardiac arrhythmias. Simultaneously, there is a need to address the mechanical function of the heart using models of the biophysics and biochemistry of molecular motors to predict the three-dimensional mechanical performance of the whole heart (Vetter and McCulloch, 1998). Related questions are how mathematical models can help improve the diagnosis and treatment of cardiac mechanical dysfunction during disease, especially congestive heart failure, and how to elucidate the mechanisms by which mechanical factors can regulate cardiac remodeling in vivo.

Integrative computational modeling of the heart has a long history dating to Laplace. Laplace's law provides an explanation for the fact that a dilated heart must create a larger wall tension in order to create the normal pressure, giving a theoretical basis for the surgical procedure of ventricular remodeling. The first cardiac myocyte ionic models were published in Noble (1962), followed by Moe's cellular automata model of atrial fibrillation in 1964 (Moe et al., 1964). Crossbridge and continuum models of ventricular mechanics started to appear in 1970. Today, an established multidisciplinary community of mathematicians, bioengineers, biophysicists, and physiologists is working on the experimental, theoretical, and computational challenges associated with formulating, implementing, and validating predictive models that integrate functionally across interacting cellular processes such as electrical excitation, mechanical contraction, and energy metabolism, and structurally across scales of biological organization from molecule to organ and system (McCulloch et al., 1998). Many in this community have advocated ambitious multicenter programs under banners such as the Cardiome Project, headed by A.D. McCulloch at the University of California at San Diego. Several large sponsored collaborations are under way (McCulloch et al., 1998; McCulloch and Huber, 2002).

However, in spite of the growing sophistication of these integrative modeling efforts, the investigators are the first to point out the manifest weaknesses and shortcomings. While excellent progress has been made in applying cellular system models of action potentials or contractile processes to three-dimensional continuum models of impulse propagation or ventricular pumping, multiscale electromechanical models are in their infancy and will require further development before they can provide the insight that is needed.

Another frontier is the development of models of the metabolic and neurohormonal (cell signaling) mechanisms that regulate excitation and contraction and their interactions (Saucerman and McCulloch, 2004). Finally, the application of integrative models to understanding the pathogenesis of genetic and acquired heart diseases and identifying new therapeutic targets is an emerging and timely field (Sussman et al., 2002).

CIRCULATORY PHYSIOLOGY

The function of the systemic circulatory system is to distribute and remove materials and heat as needed throughout the body. Transport is achieved by convection in the blood and diffusive exchange with surrounding tissue (Pittman, 2000). Because diffusion is effective only over short distances, blood must be brought close to every point in every tissue. To make this possible, the peripheral circulation consists of a highly branched system of blood vessels containing more than 109 segments ranging in diameter from about 1 cm down to a few microns. The set of vessels of diameter about 100 microns or less is referred to as the microcirculation.

A remarkable feature of the systemic circulatory system is its ability to adjust to short- and long-term changes in local functional requirements. This is achieved by a combination of central and local mechanisms. Short-term local control of blood flow is accomplished when vessels change their diameters by contracting and relaxing vascular smooth muscle. Longer-term changes in needs are met by structural changes, including changes in wall thickness and diameter and the addition of new vessels (angiogenesis). Many of these changes are driven primarily by responses to local stimuli, without central control. The peripheral circulation can therefore be considered as a highly distributed adaptive system. Understanding this system has important implications both for normal physiological processes and for many diseases, including heart disease, hypertension, and cancer.

Important areas of research are blood flow and mass transport in the microcirculation; short-term regulation of blood flow, including vascular smooth-muscle behavior; and structural adaptation of blood vessels, including angiogenesis. Mathematical and computational approaches can make important contributions in all of these areas. Continuum and multiphase models can be applied to study blood flow. Simulations of mass and heat transport also typically require solution of nonlinear partial differential equations. Consideration of network properties is also critical to understanding short- and long-term control of blood flow (Segal,

2000; Secomb and Pries, 2002). The network can be regarded as a dynamic system in which the properties of each segment (diameter, etc.) evolve with time (Segal, 2000; Ursino, 2003; Zakrzewicz et al., 2002). Simulations of angiogenesis can use a variety of approaches, including deterministic and stochastic models and cellular automata.

RESPIRATORY PHYSIOLOGY

As with many organ-level pathologies, the past decade has seen an explosion in probing lung pathology from the bottom up (e.g., genetic or cellular defects or manipulations initiating the processes that alter airway and tissue properties) as well as from the top down (e.g., advances in imaging and function measurements that provide insight on the level and extent of functional degradation). However, the chasm that remains between the two approaches must be bridged in a manner that can more effectively guide therapeutic targets and assessment. Past and even current experimental and modeling research focuses either on a specific level of lung structure—for example, on a single airway, the airway wall, tissue rheology, airway smooth muscle, or even airway smooth muscle and alveolar cell—or on function at a gross level—for example, whole-lung mechanical properties and indices of ventilation distribution. What is missing is the capacity to integrate how all the components in the lung work together to achieve dynamic function and how degradation in specific components or combinations of components might impact function. Examples of lung pathologies in need of a more comprehensive understanding of how integrated structures lead to function include asthma, adult respiratory distress syndrome, and emphysema.

Computational modeling promises a new era in the fundamental understanding of how lung morphometry and biomechanical/biomaterial properties impact lung function. With continuous improvement in imaging modalities, it is becoming increasingly possible to establish precise physical locations and degrees of structural or functional defects in the lung during disease. Such data will provide a foundation for addressing how explicit defects of biological components, processes, and structure at specific anatomic locations alter function. Computational power now permits the development of models that are closer anatomic replicas of a real lung, while incorporating the fundamental biophysical properties and relations for each component of each airway. Rational and efficient disease management could be enhanced by understanding or predicting how alterations in the individual components of lung structure and properties impact the emergent lung function.

A holy grail is a so-called in silico lung, which would reflect a personalized condition and enable simulated treatments to be performed and evaluated. Such a virtual lung would lead to the generation or rejection of specific treatment hypotheses, in turn leading to more scientific and financially cost-effective experiments or technology development. While a multiscale and personalized modeling approach has emerged for other physiological systems (e.g., cardiovascular), it remains in its infancy for the lung.

INFORMATION PROCESSING

A system physiology approach is also needed for information processing in the visual system. The traditional feedforward model of the visual system invokes a sequence of processing stages, beginning with the relay of retinal input to neurons in the primary visual cortex (V1) via the lateral geniculate nucleus (LGN) and subsequent higher-order processing through a hierarchy of cortical areas. According to this model, neurons at each successive stage process inputs from increasingly larger regions of space and code for increasingly more complex aspects of visual stimuli. The selectivity of a neuron to a given stimulus parameter (e.g., orientation, color, depth) is assumed to result from the ordered convergence of afferents from the lower stages. Although feedforward models can perform a surprising number of object-recognition tasks in some simple environments, they perform badly in many situations that are simple for human vision—for example, where an object might be partially masked or occluded by other objects. It has become clear that more complex forms of visual information processing require global-to-local interactions, both within a given stage and between different stages of the visual hierarchy. Long-range horizontal connections provide an anatomical substrate for the former, whereas feedforward and feedback connections provide a substrate for the latter. One of the major outstanding theoretical challenges is to bridge the gap between the systems physiology of vision, characterized by spatiotemporal dynamics at multiple scales (synapses, neurons, networks), and the computational/information theoretic aspects of vision (neural code, statistics of natural scenes, redundancy) (Barlow, 1961; Laughlin, 1981; Atick et al., 1992).

In a similar way, the human auditory system from the inner ear to the auditory cortex is a complex multilevel pathway of sound information processing (Dallos et al., 1990). One of the early stages of sound processing occurs in the cochlea, where the vibration pattern of the basilar membrane encodes the acoustic characteristics of incoming sound signals. Though well-known partial differential equations in classical mechanics provide a solid foundation for describing these mechanical activities, additional nonlinearities and active behaviors must be modeled to capture nonlinear responses such as tonal suppressions and the observed frequency selectivity. The next level of information processing occurs in as many as 30,000 nerve fibers connecting the inner ear to the brain. Nonlinearities are associated with peripheral auditory neurons when the hair cell converts sound signals from mechanical to neural representation. It is now well known that outer hair cells of the cochlea play an active role in increasing the sensitivity and dynamic range of the ear. In addition, the frequency distribution of sound is maintained by the wave patterns on the basilar membrane and is preserved along the fibers, resulting in an organization of frequency responses in the auditory cortex of the brain whereby different tone frequencies are transmitted separately along different parts of the structure. An additional challenge regarding processing in the auditory system is to account for the extremely fast temporal resolution of hearing, which is at the timescale of microseconds rather than the typical millisecond timescale of individual neurons.

Although mathematical models exist for many levels of visual and auditory processing, a better understanding of the connections within the systems will depend on progress in physiological experiments as well as theoretical advances to connect these individual levels. Moreover, many of the neural models take the form of integro-differential equations, in which the interaction kernel represents the spatial distribution of synaptic weights. Such equations are much less well understood than the more familiar partial differential equations of reaction diffusion systems.

ENDOCRINE PHYSIOLOGY

Mathematical modeling has led to an improved understanding of several important endocrine processes (Bertram and Sherman, 2004; Kukkonen et al., 2001; Mosekilde et al., 2001). However, there are numerous areas of endocrine physiology and associated pathologies that are in need of a more integrated approach. Consider, for instance, diabetes. It is well appreciated within the diabetes research community that diabetes is a multifactorial disease that involves the interaction over disparate spatial scales (from genes to cells to organs to the whole body) and timescales (from milliseconds to decades) (Porksen et al., 1997; Sedaghat et al., 2002; Smolen et al., 2001; Topp et al., 2000). In addition, genetic, metabolic, and ionic events all have to be integrated to achieve a workable understanding of normal and pathological regulation (Bergmann, 1989; Cobelli et al., 1998; Tornheim, 1997).

Some diabetes-related questions that could be explored through a system physiology approach include these: How is weight regulated? Is there such a thing as a set point for weight? Why is it easier to gain weight than to lose it? Why is insulin resistance associated with inflammation, hypertension, and a high LDL/HDL cholesterol ratio? How do insulin resistance and beta-cell failure interact to produce a global failure of regulation? Can we go beyond descriptive diagrams of hormone/peptide interaction networks to predictive models? How do ionic and metabolic oscillations articulate to produce pulsatile insulin secretion? How is the renin-angiotensin system regulated genetically, and what are the genetic factors underlying high blood pressure? These questions invite mathematical modeling and simulations, some of which are currently taking place.

MORPHOGENESIS AND PATTERN FORMATION

The combination of developmental genetics with rapidly advancing imaging and transcriptional profiling technologies promises a golden age for the modeling and computational analysis of developing systems. The main efforts for modeling and computational analysis can be subdivided into three groups: (1) analysis and synthesis of genetic and imaging data, with the main goal of formulating realistic models, (2) formulation of models that reflect the complexity of developing tissues, and (3) analysis of these models and their testing in direct genetic experiments.

Models of developmental pattern formation are necessarily spatially distributed, dynamic, and multivariable. All of these aspects can be now explored experimentally, opening the door to a mutually beneficial interplay between modelers and experimentalists. The expression of tens to thousands of genes in any given context can be visualized by multicolor in situ hybridizations; antibody stainings; spatially resolved, quantitative, real-time polymerase chain reactions (PCRs); or microarray experiments (Tomancak et al., 2002; Fraser and Marcotte, 2004). Few of the current techniques for gene-expression analysis in development are real-time, so the multivariable dynamics of a system have to be pieced together from a number of still shots taken from different embryos. Data normalization and image processing techniques, such as morphing, can be used to construct the spatiotemporal atlases of gene expression or pathway activity (Pereanu and Hartenstein, 2004). Recently, this approach was successfully used to develop a comprehensive multivariable and dynamic picture of gap gene expression in fruit fly embryogenesis (Kozlov et al., 2002; Jaeger et al., 2004). The spatiotemporal information about gene expression or pathway activity can be integrated, again through morphing, with the results of real-time microscopic analysis of the morphological changes in the developing system (Huisken et al., 2004; Kosman et al., 2004).

Atlases of gene expression and pathway activity directly lead to models. At the simplest level, correlation between expression patterns of multiple genes can be used to formulate Boolean or Bayesian models of gene regulation. Systematic methods for formulating such models from data must be developed, along with the computational techniques for their analysis, with an emphasis on spatially distributed systems (Friedman, 2004; Nachman et al., 2004). At the next level of complexity, multivariable spatial data can be used to fit parameters in the continuous-time dynamic models—for example, in the form of reaction-diffusion equations. This requires robust numerical methods for parameter estimation based on spatially resolved and dynamic data. Reinitz and co-workers have used stochastic optimization to fit parameters in the reaction-diffusion model of gene regulation in the early *Drosophila* embryo (Jaeger et al., 2004). The estimated parameters can be used to interpret the dynamics of genetic interactions in development.

Gene expression patterns in developing tissues can be very fine-grained, with the characteristic domains of gene expression spanning only a few cell diameters. Such patterns cannot be captured with the continuum models traditionally used to model developmental patterning (Murray, 1993). Models of developing tissues must account for cell-cell interactions by both localized and spreading signals and for the dynamics of gene expression mediated by extracellular signals (Monk, 2000; Shvartsman et al., 2002; Eldar et al., 2003). An important modeling direction involves incorporating cell-level models into the descriptions of multicellular systems and tissues (Pribyl et al., 2003).

A major challenge for the development of truly predictive pattern formation models lies in choosing the appropriate modeling formalism for describing the regulatory patterns of gene expression. Indeed, the expression of a single gene can be a highly complex function of extracellular conditions (Yuh et al., 1998; Davidson, 2001; Setty et al., 2003). Despite this complexity, it is worthwhile to explore the utility of simple logic and switchlike models for modeling gene expression in developing tissues (Thieffry and Sanchez, 2003). The most productive approach to modeling is likely to be hybrid, with threshold functions that couple extracellular signals to gene expression in individual cells. Numerical techniques for the analysis of such models are being developed (Ghosh and Tomlin, 2001).

Analysis of robustness is crucial in the evaluation of mathematical and computational models of development (von Dassow et al., 2000; Eldar et al., 2004). Indeed, developing systems can frequently tolerate gene dose reductions due to heterozygocity, so they are robust to twofold changes in the developmental parameters. This experimental observation can be used to rule out models and mechanisms that require fine-tuning of parameters. In fact, the robustness that is so common in biological systems—and which is seen as a relative insensitivity to variations in parameters—suggests that a model that requires fine-tuning may be overlooking a layer of regulation in the system. In connection with this, modeling standards for robustness analysis must be developed. Current approaches are based on random sampling of system parameters (von Dassow et al., 2000; Meir et al., 2002; Eldar et al., 2003). More sophisticated methods for parameter sampling and statistical verification of results of random parameter sampling must be developed.

The amazing robustness of developing systems, e.g., the stability of the morphologies of eggshells or wings, is contrasted with large interspecies variations. Since the time of Turing, the nonlinear instabilities induced by variations of system parameters were considered one of the mechanisms for generating increasingly complex patterns and morphologies (Turing, 1952; Meinhardt and Gierer, 2000). While true in physicochemical systems, this hypothesis still awaits its experimental verification in developing systems. This verification requires the identification of appropriate experiments where system parameters can be varied and the effects of these variations on gene expression patterns and the emerging morphologies can be examined. Model organisms, such as fruit flies and worms, can be used to generate genetic backgrounds with controlled levels of gene expression. The design of such genetic experiments can be model-based in the sense that nonlinear analysis of the model can suggest the most critical genetic perturbations (Shvartsman et al., 2002; Nakamura and Matsuno, 2003). In addition, analysis of the results of genetic experiments (e.g., the effect of overexpressing a gene on the shape of a wing) can be accelerated by the development of new image analysis and pattern recognition tools for rapid phenotyping. For instance, when studying wing development, tools for the rapid detection of morphological changes in a large number of fruit fly cells, embryos, or wings would be desirable (Myasnikova et al., 2001; Houle et al., 2003; Kiger et al., 2003).

LOCOMOTION

An important area of mathematical analysis at both the cellular and organismal level is the study of locomotion. Much productive research has been carried out on locomotion at many scales, ranging from the mechanical repositioning of subcellular organelles to the gaits of running animals. The subject is too large to review comprehensively in this report. Instead, the committee simply cites examples that illustrate the breadth of important problems on which progress has been made.

At the subcellular level, molecular motors based on actin and tubulin polymerization are of central importance in many basic cellular processes, including chromosome segregation during cell division, cell motility, muscle contraction, and intracellular transport of organelles (for a review, see Mogilner and Oster, 2003). Although bacteria themselves lack systems based on actin and tubulin, some pathogens have evolved systems for utilizing the molecular motors of their hosts for propulsion. A dramatic example is *Lysteria monocytogenes*: The relatively simple mechanical system through which *Lysteria* moves when in contact with mammalian cells lends itself well to both detailed experimental characterization and mathematical modeling (Alberts and Odell, 2004). Rotary motors, exemplified by those that drive flagellar motion in *E. coli,* have also received extensive attention (Coombs et al., 2002; for a review, see Oster and Wang, 2003).

At an organismal level, mathematical analysis has illuminated basic mechanisms of insect flight (Combes and Daniel, 2003; Miller and Peskin, 2005). A distinctive characteristic of these studies—and those of locomotion in general—has been the close interplay between experiment and theory. This interplay has long been evident even in the analysis of locomotion at larger spatial scales. Examples include the swimming motions of lampreys (Cohen et al., 1992; Lighthill, 1995) and the gaits of quadrupeds (Buono and Golubitsky, 2001). Indeed, most biomechanical processes would be difficult to study effectively without a close connection between theory and experiment.

CANCER

While cancer can be studied at the genetic and cellular levels, it is not until it is understood at the tumor level that its intrinsic cancerous behavior can be recognized. It follows that the outcome of chemotherapy cannot be understood without understanding the effects of spatial organization and intercellular communication on the dynamics of tumor development. The dynamic interplay of several biological factors determines the response of a cell to therapy and, ultimately, the outcome of chemotherapy. The key issues are (1) delivery of therapy to target tumor cells, (2) mechanisms of drug action, (3) growth and differentiation of cell populations, and (4) development of resistance.

Delivery of Therapy to Target Tumor Cells

Over 80 percent of human cancers are solid tumors. Presentation of a drug to cells in a solid tumor and the accumulation and retention of a drug in tumor cells depend on how the drug is delivered, the ability of the drug to diffuse through the interstices, and the binding of the drug to intracellular macromolecules. Some of these factors also depend on time and drug concentration. For example, the interstices, which determine the porosity and therefore the diffusion coefficient, might be expanded as a result of drug-induced apoptosis. Mathematical models depicting how these processes affect drug delivery to tumor cells could suggest the treatment regimens that will result in the most effective drug concentration and residence time in the target sites.

Mechanisms of Drug Action

Most anticancer drugs act on specific molecular targets, often molecules that are involved in the regulation of cell growth, cell differentiation, and cell death. Mathematical models to link the effective drug concentration in the tumor cells with the molecular targets, in a time- and concentration-dependent manner, are needed to improve the understanding of drug-target interaction (see Panetta et al., 2000; McDougall et al., 2002; http://calvino.polito.it/~biomat/).

Growth and Differentiation of Cell Populations

Efforts here involve the modeling of growth and differentiation of laboratory cell populations, of populations of normal cells, and of cells in tumors. Precise mathematical models exist for the processes of haemopoiesis (blood cell production) and self-renewal of colon epithelium. The mathematical tools used include stochastic processes (which are useful when describing small colonies or early stages of cancer), particularly branching processes, nonlinear ordinary differential equations (which are useful for modeling feedbacks of cell-production systems), and integral equations and partial differential equations (which are useful for modeling heterogeneous populations). The challenges involve integrating newly described genetic and molecular mechanisms into the models of proliferation, mathematically modeling the geometric growth of tumors in various phases (prevascular, vascular, anoxic), and modeling the heterogeneity of tumor populations. The mathematical tools needed include partial differential equations with free boundary conditions, bifurcation in systems of many nonlinear ordinary differential equations, and branching processes with infinite-type space.

Development of Resistance

Cancer cells are genetically unstable and can acquire genetic and phenotypic changes that permit them to escape cytotoxic insults. Development of drug resistance is common, and it is a major problem in cancer chemotherapy. Development of drug resistance is often a function of the frequency, intensity, and duration of drug exposure, as well as the chronological age of the cells. These biological parameters can be described in mathematical terms.

The modeling and optimization of chemotherapy protocols is an area of potentially great practical importance. Classical models involve populations of normal and cancer cells described as systems of ordinary differential equations with control terms representing treatment intervention. The most common classical approach involves defining a performance index that summarizes the efficiency of the therapy and the damage done to normal (noncancer) cells and using methods of control theory to find the best value of the index. These models had a good deal of appeal in the early days of chemotherapy, when the complexity of tumor cell populations was not entirely appreciated. There also exist models that take into account emerging resistance (like the Coldman-Goldie clonal resistance model) and heterogeneity (like gene amplification), but they are based on unrealistic biological hypotheses. Challenges for the field involve the development of more realistic models of drug action and cell proliferation and heterogeneity as well as new methods for parameter estimation.

IN VIVO DYNAMICS OF THE HIV-1 INFECTION

Mathematical models of HIV infection and treatment have provided quantitative insights into the main biological processes that underlie HIV pathogenesis and helped establish the treatment of patients with combination therapy.[1] This in turn has changed HIV from a fatal disease to a treatable one. The models successfully describe the changes in viral load in patients under therapy and have yielded estimates of how rapidly HIV is produced and cleared in vivo, how long HIV-infected cells survive while producing HIV, and how fast HIV mutates and evolves drug resistance. They have also provided clues to the T-cell depletion that characterizes AIDS. The models also allow the rapid screening of antiviral drug candidates for potency in vivo, hastening the introduction of new antiretroviral therapies.

HIV on average takes about 10 years to advance from initial infection to immune dysfunction (or AIDS). During this period the amount of virus measured in a person's blood hardly changes. Because of this slow progression and the unchanging level of virus, it was initially thought that the infection was slow. It was unclear if treating the disease early, when symptoms were not apparent, was worthwhile. Recognizing that constant levels of virus meant that the rates of viral production and clearance were in balance but not necessarily slow, Alan Perelson and David Ho (Perelson et al., 1996) used experimental drug therapy to perturb the viral steady state. Mathematically modeling the response to this perturbation using a system of ordinary differential equations that kept track of the concentra-

---

[1]The material in this section was generously contributed by Alan Perelson.

tions of infected cells and HIV and fitting the experimental data to the model revealed a plethora of new features about HIV infection. After therapy was initiated, levels of HIV RNA (a surrogate for virus) fell 10- to 100-fold in the first week or two of therapy. This suggested that HIV has a half-life of 1 or 2 days, so to maintain the pretherapy constant level of virus requires enormous virus production—in fact the amount of virus in the body must double every 1 or 2 days. Detailed analysis showed that this viral decay was governed by two processes: the clearance of free virus particles and the loss of productively infected cells. From this rapid clearance of virus one could compute that at steady state, $\sim 10^{10}$ virions are produced daily and, given the mutation rate of HIV, that each single and most double mutations of the HIV genome are produced daily. Thus, effective drug therapy would require drug combinations that can sustain at least three mutations before resistance arises, and this engendered the idea of triple combination therapy. Other analyses showed that the slope of viral decay was proportional to the drug combination's antiviral efficacy, providing a means of comparing therapies.

Following the rapid 1-2 week, first-phase loss, the rate of HIV RNA decline slows. Models of this second phase of decline, when fitted to the kinetic data, suggested that a small fraction of infected cells might live for a period of weeks while infected. Following on the success of these joint modeling and experimental efforts, many similar studies were undertaken that revealed a fourth, much longer timescale, between 6 and 44 months, for the decay of latently infected cells. Latently infected cells, which harbor the HIV genome but do not produce virus, can hide from the immune system and reignite infection when the cells become stimulated into proliferation. Clearing latently infected cells is one of the last remaining obstacles to eradicating HIV from the body.

The modeling of the HIV virus is but one example of the extensive contributions of the mathematical sciences to immunology and epidemiology. Many exciting opportunities remain.

FUTURE DIRECTIONS

The examples described above briefly illustrate the broad challenge and opportunities for mathematical modeling and simulation in system physiology. The use of mathematical models to describe processes in system physiology will improve our understanding of the dynamic interplay between those processes and ultimately aid in the translation of basic science findings to clinical application. At the same time, these mathematical investigations will undoubtedly lead to new mathematical problems and to new mathematical and computational methods with application in many other areas of science.

The multiscale issues of modeling at the organismal level will continue to pose what is perhaps the ultimate challenge in mathematical applications to biology. In organisms, there are often direct, immediate consequences of molecular processes: Cause-and-effect cascades explode from a scale of angstroms and picoseconds to one of meters and milliseconds, seconds, hours, or years. When integrating knowledge of organisms into the analysis of populations, which are the focus of the next chapter, it will often be possible to treat the individual organisms as homogeneous entities. However, knowledge of molecular and cellular processes will need to be taken into direct account in many models of organismal function. This goal will pose continuing, monumental challenges for scientists and mathematicians alike.

REFERENCES

Alberts, J.B., and G.M. Odell. 2004. In silico reconstitution of Listeria propulsion exhibits nano-saltation. *PLoS Biol.* 2(12): e412.

Atick, J.J., Z. Li, and A.N. Redlich. 1992. Understanding retinal color coding from first principles. *Neural Comput.* 4: 559-572.

Barlow, H.B. 1961. Possible principles underlying the transformation of sensory messages. Pp. 217-234 in *Sensory Communication*. W. Rosenblith, ed. Cambridge, Mass.: MIT Press.

Bergman, R.N. 1989. Lilly lecture 1989: Toward physiological understanding of glucose tolerance: Minimal-model approach. *Diabetes* 38(12): 1512-1527.

Bertram, R., and A. Sherman. 2004. A calcium-based phantom bursting model for pancreatic islets. *Bull. Math. Biol.* 66(5): 1313-1344.

Buono, P.L., and M. Golubitsky. 2001. Models of central pattern generators for quadruped locomotion. I. Primary gaits. *J. Math. Biol.* 42(4): 291-326.

CAST Investigators. 1989. Preliminary report: Effect of encainide and flecainide on mortality in a randomized trial of arrhythmia suppression after myocardial infarction. *N. Engl. J. Med.* 321: 407-412.

Cobelli, C., F. Bettini, A. Caumo, and M.J. Quon. 1998. Overestimation of minimal model glucose effectiveness in presence of insulin response is due to undermodeling. *Am. J. Physiol.* 275(6 Pt 1): E1031- E1036.

Cohen, A.H., G.B. Ermentrout, T. Kiemel, N. Kopell, K.A. Sigvardt, and T.L. Williams. 1992. Modelling of intersegmental coordination in the lamprey central pattern generator for locomotion. *Trends Neurosci.* 15(11): 434-438.

Combes, S.A., and T.L. Daniel. 2003. Flexural stiffness in insect wings. II. Spatial distribution and dynamic wing bending. *J. Exp. Biol.* 206(Pt 17): 2989-2997.

Coombs, D., G. Huber, J.O. Kessler, and R.E. Goldstein. 2002. Periodic chirality transformations propagating on bacterial flagella. *Phys. Rev. Lett.* 89(11): 118102.

Davidson, E.H. 2001. *Genomic Regulatory System*. San Diego, Calif.: Academic Press.

Dallos, P., C.D. Geisler, J.W. Matthews, M.A. Ruggero, and C.R. Steele, eds. 1990. *The Mechanics and Biophysics of Hearing. Lecture Notes in Biomathematics 87*. Heidelberg: Springer-Verlag.

Eldar, A., D. Rosin, B.Z. Shilo, and N. Barkai. 2003. Self-enhanced ligand degradation underlies robustness of morphogen gradients. *Dev. Cell* 5(4): 635-646.

Eldar, A., B.Z. Shilo, and N. Barkai. 2004. Elucidating mechanisms underlying robustness of morphogen gradients. *Curr. Opin. Genet. Dev.* 14(4): 435-439.

Fraser, A.G., and E.M. Marcotte. 2004. Development through the eyes of functional genomics. *Curr. Opin. Genet. Dev.* 14(4): 328-335.

Friedman, N. 2004. Inferring cellular networks using probabilistic graphical models. *Science* 303(5659): 799-805.

Ghosh, R., and C. Tomlin. 2001. Lateral inhibition through delta-notch signaling: A piecewise affine hybrid model. Pp. 232-246 in *Hybrid Systems: Computation and Control, Lecture Notes in Computer Science 2034*. M.D. Di Benedetto and A.L. Sangiovanni-Vincentelli, eds. New York, N.Y.: Springer-Verlag.

Houle, D., J. Mezey, P. Galpern, and A. Carter. 2003. Automated measurement of Drosophila wings. *BMC Evol. Biol.* 3(1): 25.

Huisken, J., J. Swoger, F. Del Bene, J. Wittbrodt, and E.H. Stelzer. 2004. Optical sectioning deep inside live embryos by selective plane illumination microscopy. *Science* 305(5686): 1007-1009.

Igoshin, O.A., and Oster G. 2004. Rippling of myxobacteria. *Math. Biosci.* 188: 221-233.

Jaeger, J., S. Surkova, M. Blagov, H. Janssens, D. Kosman, K.N. Kozlov, Manu, E. Myasnikova, C.E. Vanario-Alonso, M. Samsonova, D.H. Sharp, and J. Reinitz. 2004. Dynamic control of positional information in the early Drosophila embryo. *Nature* 430(6997): 368-371.

Jafri, M.S., J.J. Rice, and R.L. Winslow. 1998. Cardiac calcium dynamics: The roles of ryanodine receptor adaptation and sarcoplasmic reticulum load. *Biophys. J.* 74: 1149-1168.

Kenknight, B., B. Jones, A. Thomas, and D. Lang. 1996. Technological advances in implantable cardioverter-defibrillators before the year 2000 and beyond. *Am. J. Cardiol.* 78: 108.

Kiger, A., B. Baum, S. Jones, M.R. Jones, A. Coulson, C. Echeverri, and N. Perrimon. 2003. A functional genomic analysis of cell morphology using RNA interference. *J. Biol.* 2(4): 27.

Kosman, D., C.M. Mizutani, D. Lemons, W.G. Cox, W. McGinnis, and E. Bier. 2004. Multiplex detection of RNA expression in Drosophila embryos. *Science* 305(5685): 846.

Kozlov, K., E. Myasnikova, A. Pisarev, M. Samsonova, and J. Reinitz. 2002. A method for two-dimensional registration and construction of the two-dimensional atlas of gene expression patterns in situ. *In Silico Biol.* 2(2): 125-141.

Kukkonen, J.P., J. Nasman, and A.E. Akerman. 2001. Modelling of promiscuous receptor-Gi/Gs-protein coupling and effector response. *Trends Pharmacol. Sci.* 22(12): 616-622.

Laughlin, S.B. 1981. A simple coding procedure enhances a neuron's information capacity. *Z. Naturforsch.* 36c: 910-912.

Lighthill, J. 1995. The role of the lateral line in active drag reduction by clupeoid fishes. *Symp. Soc. Exp. Biol.* 49: 35-48.

Luo, C.H., and Y.A. Rudy. 1994. A dynamic model of the cardiac ventricular action potential, I: Simulations of ionic currents and concentration changes. *Circ. Res.* 74: 1071-1096.

McCulloch, A.D., and G. Huber. 2002. Integrative biological modelling in silico. Pp. 4-25 in *'In Silico' Simulation of Biological Processes No. 247*. Novartis Foundation Symposium. G. Bock and J.A. Goode, eds. Chichester, U.K.: John Wiley & Sons Ltd.

McCulloch, A.D., J.B. Bassingthwaighte, P.J. Hunter, and D. Noble. 1998. Computational biology of the heart: From structure to function. *Progr. Biophys. Mol. Biol.* 69: 153-155.

McDougall, S.R., A.R.A. Anderson, M.A.J. Chaplain, and J.A. Sherratt. 2002. Mathematical modelling of flow through vascular networks: Implications for tumour-induced angiogenesis and chemotherapy strategies. *Bull. Math. Biol.* 64: 673-702.

Meinhardt, H., and M. Gierer. 2000. Pattern formation by local self-activation and lateral self-inhibition. *Bioessays* 22(8): 753-760.

Meir, E., G. von Dassow, E. Munro, and G.M. Odell. 2002. Robustness, flexibility, and the role of lateral inhibition in the neurogenic network. *Curr. Biol.* 12(10): 778-786.

Miller, L.A., and C.S. Peskin. 2005. A computational fluid dynamics of 'clap and fling' in the smallest insects. *J. Exp. Biol.* 208(Pt 2): 195-212.

Moe, G.K., W.C. Rheinbolt, and J. Abildskov. 1964. A computer model of atrial fibrillation. *Am. Heart J.* 67: 200-220.

Mogilner, A., and G. Oster. 2003. Polymer motors: Pushing out the front and pulling up the back. *Curr. Biol.* 13(18): R721-R733.

Monk, N.A.M. 2000. Elegant hypothesis and inelegant fact in developmental biology. *Endeavour* 24(4): 170-173.

Morgenroth, E., H.J. Eberl, M.C. van Loosdrecht, D.R. Noguera, G.E. Pizarro, C. Picioreanu, B.E. Rittmann, A.O. Schwarz, and O. Wanner. 2004. Comparing biofilm models for a single species biofilm system. *Water Sci. Technol.* 49(11-12): 145-154.

Mosekilde, E., B. Lading, S. Yanchuk, and Y. Maistrenko. 2001. Bifurcation structure of a model of bursting pancreatic cells. *Biosystems* 63(1-3): 3-13.

Murray, J.D. 1993. *Mathematical Biology*. New York, N.Y.: Springer-Verlag.

Myasnikova, E., A. Samsonova, K. Kozlov, M. Samsonova, and J. Reinitz. 2001. Registration of the expression patterns of Drosophila segmentation genes by two independent methods. *Bioinformatics* 17(1): 3-12.

Nachman, I., A. Regev, and N. Friedman. 2004. Inferring quantitative models of regulatory networks from expression data. *Bioinformatics* 20(Suppl 1): I248-I256.

Nakamura, Y., and K. Matsuno. 2003. Species-specific activation of EGF receptor signaling underlies evolutionary diversity in the dorsal appendage number of the genus Drosophila eggshells. *Mech. Dev.* 120(8): 897-907.

Noble, D. 1962. A modification of the Hodgkin-Huxley equations applicable to Purkinje fiber action and pacemaker potential. *J. Physiol.* 160: 317-352.

Oster, G., and H. Wang. 2003. Rotary protein motors. *Trends Cell. Biol.* 13(3): 114-121.

Panetta, J.C., M.A.J. Chaplain, and D. Cameron. 2000. Modelling the effects of Paclitaxel and Cisplatin on breast and ovarian cancer. *J. Theor. Med.* 3: 11-23.

Pereanu, W., and V. Hartenstein. 2004. Digital three-dimensional models of Drosophila development. *Curr. Opin. Genet. Dev.* 14(4): 382-391.

Perelson, A.S., A.U. Neumann, M. Markowitz, J.M. Leonard, and D.D. Ho. 1996. HIV-1 dynamics in vivo: Virion clearance rate, infected cell life-span, and viral generation time. *Science* 271(5255): 1582-1586.

Pittman, R.N. 2000. Oxygen supply to contracting skeletal muscle at the microcirculatory level: Diffusion vs. convection. *Acta Physiol. Scand.* 168: 593-602.

Porksen, N., B. Nyholm, J.D. Veldhuis, P.C. Butler, and O. Schmitz. 1997. In humans at least 75% of insulin secretion arises from punctuated insulin secretory bursts. *Am. J. Physiol.* 273(5 Pt 1): E908-E914.

Pribyl, M., C.B. Muratov, and S.Y. Shvartsman. 2003. Discrete models of autocrine signaling in epithelial layers. *Biophys. J.* 84(6): 3624-3635.

Puglisi, J.L., and D.M. Bers. 2001. LabHEART: An interactive computer model of rabbit ventricular myocyte ion channels and Ca transport. *Am. J. Physiol.* 281: C2049-C2060.

Saucerman, J.J., and A.D. McCulloch. 2004. Mechanistic systems models of cell signaling networks: A case study of myocyte adrenergic regulation. *Prog. Biophys. Mol. Biol.* 85(2/3): 261-278.

Secomb, T.W., and A.R. Pries. 2002. Information transfer in microvascular networks. *Microcirculation* 9: 377-387.

Sedaghat, A.R., A. Sherman, and M.J. Quon. 2002. A mathematical model of metabolic insulin signaling pathways. *Am. J. Physiol. Endocrinol. Metab.* 283(5): E1084-E1101.

Segal, S.S. 2000. Integration of blood flow control to skeletal muscle: Key role of feed arteries. *Acta Physiol. Scand.* 168: 511-518.

Setty, Y., A.E. Mayo, M.G. Surette, and U. Alon. 2003. Detailed map of a cis-regulatory input function. *Proc. Natl. Acad. Sci. U.S.A.* 100(13): 7702-7707.

Shvartsman, S.Y., C.B. Muratov, and D.A. Lauffenburger. 2002. Modeling and computational analysis of EGF receptor-mediated cell communication in Drosophila oogenesis. *Development* 129(11): 2577-2589.

Smolen, P., D.A. Baxter, and J.H. Byrne. 2001. Modeling circadian oscillations with interlocking positive and negative feedback loops. *J. Neurosci.* 21(17): 6644-6656.

Sussman, M.A., A. McCulloch, and T.K. Borg. 2002. Dance band on the Titanic: Biomechanical signaling in cardiac hypertrophy. *Circ. Res.* 91: 888-898.

Thieffry, D., and L. Sanchez. 2003. Dynamical modelling of pattern formation during embryonic development. *Curr. Opin. Genet. Dev.* 13(4): 326-330.

Tomancak, P., A. Beaton, R. Weiszmann, E. Kwan, S. Shu, S.E. Lewis, S. Richards, M. Ashburner, V. Hartenstein, S.E. Celniker, and G.M. Rubin. 2002. Systematic determination of patterns of gene expression during Drosophila embryogenesis. *Genome Biol.* 3(12): Research0088.

Topp, B., K. Promislow, G. deVries, R.M. Miura, and D.T. Finegood. 2000. A model of beta-cell mass, insulin, and glucose kinetics: Pathways to diabetes. *J. Theor. Biol.* 206(4): 605-619.

Tornheim, K. 1997. Are metabolic oscillations responsible for normal oscillatory insulin secretion? *Diabetes* 46(9): 1375-1380.

Turing, A.M. 1952. The chemical basis of morphogenesis. *Phil. Trans. Roy. Soc. B* 237: 37-72.

Umeda, T., and K. Inouye. 2004. Cell sorting by differential cell motility: A model for pattern formation in Dictyostelium. *J. Theor. Biol.* 226(2): 215-224.

Ursino, M. 2003. Cerebrovascular modelling: A union of physiology, clinical medicine and biomedical engineering. Editorial. *Med. Eng. Physics* 25: 617-620.

Vetter, F.J., and A.D. McCulloch. 1998. Three-dimensional analysis of regional cardiac anatomy. *Progr. Biophys. Mol. Biol.* 69: 157-184.

von Dassow, G., E. Meir, E.M. Munro, and G.M Odell. 2000. The segment polarity network is a robust developmental module. *Nature* 406(6792): 188-192.

Waldo, A.L., A.J. Camm, H. deRuyter, P.L. Friedman, D.J. MacNeil, J.F. Pauls, B. Pitt, C.M. Pratt, P.J. Schwartz, and E.P. Veltri. 1996. Effect of d-sotalol on mortality in patients with left ventricular dysfunction after recent myocardial infarction. The SWORD investigators. Survival with oral d-sotalol. *Lancet* 348: 7-12.

Yuh, C.-H., H. Bolouri, J.M. Bower, and E.H. Davidson. 1998. Genomic cis-regulatory logic: Experimental and computational analysis of a sea urchin gene. *Science* 279(5358): 1896-1902.

Zakrzewicz, A., T.W. Secomb, and A.R. Pries. 2002. Angioadaptation: Keeping the vascular system in shape. *News Physiol. Sci.* 17: 197-201.

6

Understanding Populations

POPULATION GENETICS

From the earliest days of population genetics, mathematics has played an important role in the field. Until the 1960s, most population genetics theory focused on deductive analysis, and the models were generally focused on following the evolution of populations that were presumed to have originally been located in just one or two places. Early investigators showed how evolution would proceed under plausible models of genetic inheritance and natural selection. These analyses illuminated the dynamics of allele frequencies in populations, and they showed with what speed evolution could occur and how this speed depended on various parameters. Both deterministic models and models with random genetic drift were examined. Diffusion approximations to Markov chains were particularly important (Kimura, 1983). These diffusion processes could be analyzed by solving simple ordinary differential equations to obtain important quantities such as the probability of fixation of a new variant and the mean time for such fixation. These analyses strongly shaped our current understanding of natural selection in large, but finite, populations and guided experimental work. In recent years the emphasis has shifted from these deductive activities to inductive or retrospective approaches that address the question of what we can infer about evolutionary history and the nature of the evolutionary process from current patterns of genetic variation.

The primary goal of population and evolutionary geneticists today is to understand patterns of genetic variation within populations and patterns of genetic divergence between species. Population geneticists have asked, What are the important forces that determine the amount and nature of genetic variation in populations, the spatial distribution of this variation, the distribution of variation across the genome, and the evolutionary changes that occur over short and long timescales? The process that has shaped this variation within and between species is a complex one involving a complex genome and a complex, spatially and temporally varying environment. It is certain that stochasticity is an important aspect of the process. The rapidly growing database of DNA polymorphism and divergence studies from a variety of organisms, including humans and other primates, provide an exciting opportunity to learn about the evolutionary history of populations and the evolutionary processes that have resulted in the patterns of variation that we observe in extant populations. The difficulty is that even very simple models of this process lead to challenging mathematical problems.

Some examples of current approaches and the mathematical challenges facing us are described here. To be concrete and to avoid an overly vague description of the problems, a very specific population genetic model of sequence evolution will be described. The particular model, the Wright-Fisher model, has a long and rich history, but it is not necessarily the most realistic or tractable for every purpose, and it is only one of many models that might have been considered here.

The Wright-Fisher model assumes discrete generations (as opposed to a model with distinct age classes and overlapping generations, which would be more realistic for some populations, including humans). The focus is on a particular segment of the genome, referred to as a gene, and it is first assumed that no recombination or mutation occurs. To begin, it is assumed as well that population size (N) is constant and that there is no spatial structure. A haploid model is also assumed, which means that each individual carries just one copy of the gene. (Humans are in fact diploid, which means that each individual carries two copies of each gene, a maternal and a paternal copy.)

In the Wright-Fisher model, successive generations are produced as follows. Each of the N individuals of the offspring generation is produced by replicating, without error, the gene sequence of a randomly drawn individual of the parental generation. Each offspring individual is assumed to be generated independently from the parental population in this manner. The number of offspring of any particular individual of the parent generation is thus a random variable, being the result of N independent Bernoulli trials, with probability of success equal to $1/N$. In large populations, the number of offspring of any individual would approximately follow a Poisson distribution with mean 1. If it is supposed that the parents do not all have identical gene sequences, then their distinct gene sequences are known as haplotypes. Given the frequencies of the different haplotypes in the parental generation, the numbers of the different haplotypes in the offspring generation will be multinomially distributed. Regardless of how much variation existed in the founding population, the population under this model will eventually arrive at a state in which every individual carries the same sequence. This process of random change in the frequencies of the different haplotypes is referred to as genetic drift, and it eventually results in the population becoming monomorphic.

Next, mutation is introduced into the model. Let it be supposed that the replication process that generates an offspring copy of the gene from its parent has some error rate, so that each offspring differs from its parent at a Poisson-distributed number of sites in the gene sequence. If this model is run for many generations, the pattern of genetic variation asymptotically approaches a stationary distribution resulting from a stochastic balance between mutation, which generates variation in the population, and genetic drift, which tends to eliminate variation. Many properties of this stationary distribution are known. Also, many properties of samples drawn from this stationary distribution are known. In the model as it has been defined here, all individuals are in some sense equivalent. For example, all individuals have the same distribution of offspring number with expectation equal to 1. All the genetic variation is said to be selectively neutral, and the model is referred to as a neutral model. In generalizations of this model, some sequence variants may have a systematic tendency to produce more offspring than others, and the frequency of such variants will tend to increase. These are models of evolution with natural selection.

The Wright-Fisher neutral model is a particular case of a more general class of neutral models in which all parents are equivalent; these are referred to as "exchangeable models." In these models, the distribution of offspring number need not be Poissonian. In the limit of large populations and a low mutation rate, the models' stationary properties depend on a single compound parameter, Nu/v, where u is the mutation rate and v is the variance of offspring number. Despite the simplicity of this model, in which there is no selection, no geographic structure, no variation in population size, and no recombination, the probabilities of sample configurations of sequences under this model are difficult to calculate. Strobeck (1983) first described recursions for these probabilities for the case where only two or three haplotypes are present in a sample. The difficulty of obtaining sample configuration probabilities led to the use of summaries of the data, with an inevitable loss of information. Only in the last 10 years have full likelihood approaches been developed. Griffiths and Tavaré (1994a, 1995) were the first to find a practical method to estimate full likelihoods for this simple neutral model using a method based on importance sampling. Kuhner et al. (1995) described a Markov chain Monte Carlo (MCMC) method for obtaining quantities proportional to the sample probabilities. The main point here is that the sampling properties of sequence data under even this simplest, one-parameter, neutral model lead to recursions that are not analytically tractable. Monte Carlo methods have provided a way forward. Much of the progress in understanding these models is based on analyzing properties of the genealogical relationships of sampled copies rather than analyzing the dynamics of population frequencies of haplotypes. The population genetics theory of sample genealogies has come to be known as calescent theory. The early important work in this area was done by Watterson (1975), Kingman (1982), and Tajima (1983). (See Chapter 2 for more information on this.)

For models without recombination, it is possible to extend these Monte Carlo methods to the case in which population size is not held constant. For some special cases, it is possible to infer past population size changes (Griffiths and Tavaré, 1994b; Kuhner et al., 1995). Additional Monte Carlo methods for demographic inference using other types of genetic data (microsatellite data, for example) have also been developed (Beaumont, 1999). Much remains to be done in this area.

Models with geographic structure are more difficult. Historically, simple "island" models have been employed. These subdivide the population, but with a special structure in which each subdivision is assumed to communicate equally with all other subdivisions. More realistic stepping-stone models are more difficult, but some results are known (Durrett, 2002). Wakeley (2004) recently investigated a class of models with large numbers of subdivisions and obtained elegant results for this model. This work has capitalized on results for coalescent processes that operate on different timescales. Bayesian Monte Carlo methods have again begun to play an important role in analyzing data of several types (Pritchard et al., 2000; Beaumont, 1999).

If recombination is added to the model, the difficulties increase enormously. With recombination, each offspring produced in the model has some small probability, r, of being the product of two parent individuals, one parent contributing a part of the gene on the left and the other parent contributing the rest of the gene, the boundary between the two contributions being random. In models with recombination, complex statistical dependencies between sites arise. Sample configuration probabilities are very difficult to obtain. For a model with just two sites between which recombination can occur, a relatively simple recursion can be written down for sample probabilities (Golding, 1984). These recursions are intractable for all but very small samples, as the state space becomes enormous. For more than two sites the situation quickly gets much worse.

Griffiths and Marjoram (1996) present recursions for sample configuration probabilities under the infinite-sites version of this model with recombination. These are not analytically tractable, but Monte Carlo methods have been described for estimating these sampling probabilities under this model. However, unlike the case without recombination, it appears that these Monte Carlo methods are computationally infeasible for samples of interesting size because convergence, while difficult to assess, appears to take inordinate amounts of computer time. As a consequence, approximate methods, ad hoc methods, and methods based on summary statistics are still the rule when analyzing data from genes with recombination (Stephens, 2001). Much interest has focused on making inferences about recombination rates and gene conversion rates under models in which the rates vary across the genome (McVean et al., 2004). Improved methods could contribute to understanding the genetic mechanisms of recombination and also help in the mapping of disease genes via association studies.

There is also great interest in assessing the importance of natural selection in shaping patterns of variation within populations and the divergence between populations. Many ad hoc tests have been developed over the years (Kreitman, 2000; Bustamante et al., 2003) to explore these questions. Devising methods that make more efficient use of the data and combine information from many loci should be a priority. These inferences about selection must be made in the context of realistic models of population structure and demographic history. More realistic models of mutation and recombination are also needed, building on results shown in recent work such as (Hwang and Green, 2004; Meunier and Duret, 2004). This recent work points out how much there is still to learn about molecular evolution and how rich the mathematical models will need to be to capture it.

The focus thus far in this chapter has been on variation within species, but comparisons of sequences from different species can also be very informative, both about evolutionary relationships of species (the phylogenetic inference problem) and about evolutionary processes. Again, recent years have seen important progress on likelihood methods (Felsenstein, 1981) and, most recently, on the use of Monte Carlo approaches (Huelsenbeck and Ronquist, 2001; Wong et al., 2004). Combining between- and within-species data can be very useful, as is well exemplified by the analysis of Poisson random field models (Bustamante et al., 2003). A remaining challenge in phylogenetic inference includes the problem of performing multiple alignments and phylogenetic reconstruction simultaneously. Currently, alignment is carried out while ignoring phylogenetic relationships, and phylogenetic reconstruction is carried out only with fully aligned sequences produced prior to the reconstruction. This clearly is not an optimal solution.

Genetic data are becoming available at an ever-increasing rate. More loci, more species, and more individuals within species will be surveyed. More and more frequently, essentially complete genomes will be compared. These advances result in new opportunities and new mathematical and computational challenges. Different biological questions, different organisms, and different types of genetic markers will require somewhat different models, different methods, and different approximations. With more and richer data sets, researchers will be able to fruitfully consider somewhat more complex models, with increased demographic complexity (bottlenecks and expansions, more complex spatial structure) and increased genetic complexity (heterogeneous recombination and mutation rates), and with more complex types of natural selection (interaction between sites and spatial and temporal variation in selection coefficients). These complexities present significant mathematical challenges. Stochastic models, possibly with many parameters, and complex, nonindependent data make the computational difficulties substantial. Insight about the models and mathematical skill will be needed to make progress. While a wide range of approaches will clearly contribute to advances, the important roles that Monte Carlo methods and Bayesian approaches have recently played seem likely to continue.

ECOLOGICAL ASPECTS OF POPULATIONS

Population growth with density dependence was formulated mathematically by Verhulst (1838), who developed a number of models to investigate the consequences of deviations from unrestricted growth. One of the models, known as the logistic equation, remains a standard model for population growth. It is described by the following differential equation:

$$\frac{dN}{dt} = rN\left(1 - \frac{N}{K}\right)$$

where $N(t)$ is the population size at time t, r is the intrinsic rate of growth, and K is the carrying capacity. This model continues to serve as an important illustration of the effects of negative density dependence, as encoded by the term in parentheses. It is an example of a phenomenological model, one that is intended to embody, in a concise form, some of the observed behaviors of populations, but it has also been used as a predictive model: Based on data on the U.S. population from 1790 to 1910, Pearl and Reed

(1920) fitted the logistic equation to the observed growth of the U.S. population, estimating that it would level off at 197 million.

A discrete-time version of the logistic growth equation exhibits surprising properties, from periodic behavior to chaos. This latter behavior was introduced to ecology by May (1974, 1976). Although a known phenomenon in mathematics, the idea that deterministic models can exhibit unpredictable behavior was new to ecologists at that time and still spawns new research in ecology (Cushing et al., 2002). Although difficult to verify in nature, experimental systems have been developed to test for chaos (Costantino et al., 1997). The theoretical study of single populations in a purely ecological context remains challenging. Much research is being devoted to anthropogenic impacts on natural and managed systems. Mathematical challenges include modeling and analysis of spatial aspects, temporal and spatial variation, demographic stochasticity (in particular when dealing with small populations), and nonequilibrium dynamics. The committee highlights two areas of interest: species extinction and food supply.

The threat of species extinction arising from either habitat fragmentation or species invasion has resulted in much theoretical work. The need here is for both conceptual models, which will give us a better understanding of the underlying mechanisms, and predictive models, which can be used for management. Much of the theoretical work focuses on single-species models. On the empirical side, observational studies dominate, and data are not always unequivocal (Debinski and Holt, 2000). There are few controlled experiments of habitat fragmentation and species invasions because of the difficulties associated with these experiments.

Reliable food supply depends on the ability to manage this renewable resource. Fisheries management is an example that has enjoyed sophisticated modeling by both economists and biologists. Conceptual and highly species-specific models are widespread, but many focus only on single populations. Large, spatially explicit data sets exist, although there are troublesome uncertainties associated with some of the data. More realistic models also must deal with the uncertainties associated with management strategies and inherently stochastic processes, such as birth and death. They also need to take food web structure into account. A comprehensive model that includes visualization tools was developed by James Kitchell and co-workers for the Central North Pacific ecosystem to assess the effects of fishing on productivity (Hinke et al., 2004). Visualizations are useful to convey the impact of different management scenarios to managers. The development of such complex models requires a thorough understanding of the ecological interactions in addition to long-term data to parameterize the models and test scenarios.

These models are typically so complex that simulation is the only tool currently available for investigation.

In the two cases discussed here, the need is for models that incorporate the complex interactions of the target species with its surroundings, often in spatially heterogeneous habitats and under nonequilibrium conditions. Models that are used for management purposes often include social and economic aspects. The importance of developing methods to study nonequilibrium dynamics cannot be overemphasized (see, for example, Hastings, 2004). Some of the mathematical theory has been developed, in particular when different timescales are involved. Many ecological interactions occur far away from equilibrium, and large-scale anthropogenic perturbations, such as land use change, species invasions, or alterations of nutrient or carbon cycles, exacerbate this problem. Land use change and the invasion of exotic species often result in rapid changes and thus have the potential to move a system far from equilibrium, with consequences for both ecological and evolutionary processes (see below). Even experimental systems are probably not in equilibrium: Most field experiments are studied over only short timescales, even if the dynamics are slow.

A SYNTHESIS OF ECOLOGY AND EVOLUTION

During ecology's early years, evolutionary thinking was prevalent. However, as ecology focused more and more on abiotic and biotic causes of diversity and species abundance, evolutionary thinking became less prominent (Collins, 1986). Much of ecology now operates under the premise that ecological and evolutionary processes act on different timescales. Evolutionary processes are often thought to take hundreds of generations before their effects can be measured, whereas ecological processes often show effects after a few generations. This has led to the intellectual separation of ecology and evolution. For systems that are under strong selection, however, this may not be the case. There are classic examples, such as melanism among moths as a response to air pollution (Kettlewell, 1955) or the heavy-metal tolerance of plants (Bradshaw, 1952). As a consequence, purely ecological or purely genetic models are often inadequate when strong selection is acting (Neuhauser et al., 2003). An increasing number of studies are combining ecological and evolutionary models to meet this challenge of understanding the consequences of ecological and evolutionary forces acting on similar timescales (Antonovics, 1992; Thompson, 1999; Whitman et al., 2003).

The mathematical challenges when both ecological and evolutionary processes are considered simultaneously are numerous. First, the dimensionality increases because additional parameters must be introduced to model both ecological and evolutionary processes. Second, these processes are frequently both spatial and stochastic. The study of spatial stochastic systems is an active area of mathematical research, but at this point, only the simplest models seem to be tractable in a rigorous way. Third, the interplay between ecological and evolutionary processes is most pronounced during transient dynamics. Since there are no readily available analytical methods for nonequilibrium processes that are spatial and stochastic, most studies resort to simulations as the primary way to gain insights. The following example illustrates a biological problem and the mathematical challenges it brings.

As an example of why one might wish to couple ecological and evolutionary processes, and of the mathematical challenges that result, consider the evolution of resistance. This is of importance, for instance, in understanding the ramifications of the use of transgenic *Bt* crops, which have been engineered to express a toxin from the soil bacterium *Bacillus thuringiensis (Bt)*. Engineered versions are available for a number of crop plants, such as maize, potatoes, cotton, and soybeans. In maize, the toxin is expressed at high levels and is toxic to the European corn borer, *Ostrinia nubilalis* (Hübner), the key herbivore insect pest. An important concern is the pest's development of resistance to the toxin (Tabashnik, 1994; Gould, 1998). Current practice is to plant a "refuge" of non-*Bt* maize alongside *Bt* maize to allow sufficient numbers of susceptible European corn borers to be available as mates if resistant types emerge from the *Bt* field. To model the evolution of resistance, two things are needed: nonequilibrium models for at least two types of patches (*Bt* field and non-*Bt* field) and the ability to study the time-varying genetic composition of European corn borer populations throughout these modeled patches. One of the first models that incorporated these aspects was developed by Comins (1977). Since then, other models have been developed, and each seems to reveal additional complexities. A consistent theory is still lacking, and it will need to also take into account the community dynamics of associated enemies of the insect pest (see, for example, Neuhauser et al., 2003).

The field of population biology focuses largely on single populations. Except under controlled experimental situations, populations rarely live in isolation. Populations are typically embedded in communities, and their dynamics are strongly influenced by other members of the community. These feedbacks greatly complicate our understanding of the dynamics and present great challenges. The next chapter will discuss communities.

REFERENCES

Antonovics, J. 1992. Toward community genetics. Pp. 426-449 in *Plant Resistance to Herbivores and Pathogens: Ecology, Evolution, and Genetics*. R.S. Frite and E.L. Simms, eds. Chicago, Ill.: University of Chicago Press.

Beaumont, M. 1999. Detecting population expansion and decline using microsatellites. *Genetics* 153: 2013-2029.

Bradshaw, A.D. 1952. Populations of *Agrostis tenuis* resistant to lead and zinc poisoning. *Nature* 169: 1098-1099.

Bustamante, C.D., R. Nielsen, and D.L. Hartl. 2003. Maximum likelihood and Bayesian methods for estimating the distribution of selective effects among classes of mutations using DNA polymorphism data. *Theor. Popul. Biol.* 63(2): 91-103.

Collins, J.P. 1986. Evolutionary ecology and the use of natural selection in ecological theory. *J. Hist. Biol.* 19: 257-288.

Comins, H.N. 1977. The development of insecticide resistance in the presence of migration. *J. Theor. Biol.* 64: 177-179.

Costantino, R.F., R.A. Desharnais, J.M. Cushing, and B. Dennis. 1997. Chaotic dynamics in an insect population. *Science* 275: 389-391.

Cushing, J.M., R.F. Constantino, B. Dennis, R.A. Desharnais, and S.M. Henson. 2002. *Chaos in Ecology: Experimental Nonlinear Dynamics*. San Diego, Calif.: Academic Press.

Debinski, D.M., and R.D. Holt. 2000. A survey and overview of habitat fragmentation experiments. *Conserv. Biol.* 14: 342-355.

Durrett, R. 2002. *Probability Models for DNA Sequence Evolution*. New York, N.Y.: Springer-Verlag.

Felsenstein, J. 1981. Evolutionary trees from DNA sequences: A maximum likelihood approach. *J. Mol. Evol.* 17(6): 368-376.

Golding, G.B. 1984. The sampling distribution of linkage disequilibrium. *Genetics* 108: 257-274.

Gould, F. 1998. Sustainability of transgenic insecticidal cultivars: Integrating pest genetic and ecology. *Annu. Rev. Entomol.* 43: 701-726.

Griffiths, R.C., and S. Tavaré. 1994a. Simulating probability distributions in the coalescent. *Theor. Popul. Biol.* 46: 131-159.

Griffiths, R.C., and S. Tavaré. 1994b. Sampling theory for neutral alleles in a varying environment. *Phil. Trans. Roy. Soc. Lond. B Biol. Sci.* 344(1310): 403-410.

Griffiths, R.C., and S. Tavaré. 1995. Unrooted genealogical tree probabilities in the infinitely-many-sites model. *Math. Biosci.* 127(1): 77-98.

Griffiths, R.C., and P. Marjoram. 1996. Ancestral inference from samples of DNA sequences with recombination. *J. Comput. Biol.* 3(4): 479-502.

Hastings, A. 2004. Transients: The key to long-term ecological understanding? *Trends Ecol. Evol.* 19: 39-45.

Hinke, J.T., I.C. Kaplan, K. Aydin, G.M. Watters, R.J. Olson, and J.F Kitchell. 2004. Visualizing the food-web effects of fishing for tunas in the Pacific Ocean. *Ecol. Soc.* 9: 10.

Huelsenbeck, J.P., and F. Ronquist. 2001. MRBAYES: Bayesian inference of phylogenetic trees. *Bioinformatics* 17(8): 754-755.

Hwang, D.G., and P. Green. 2004. Bayesian Markov chain Monte Carlo sequence analysis reveals varying neutral substitution patterns in mammalian evolution. *Proc. Natl. Acad. Sci. U.S.A.* 1011(39): 13994-14001.

Kettlewell, H.B.D. 1955. Selection experiments on industrial melanism in the Lepidoptera. *Heredity* 9: 323-342.

Kimura, M. 1983. *The Neutral Theory of Molecular Evolution*. New York, N.Y.: Cambridge University Press.

Kingman, J.F.C. 1982. On the genealogy of large populations. *J. App. Prob.* 19A: 27-43.

Kreitman, M. 2000. Methods to detect selection in populations with applications to the human. *Annu. Rev. Genomics Hum. Genet.* 1: 539-559.

Kuhner, M.K., J. Yamato, and J. Felsenstein. 1995. Estimating effective population size and mutation rate from sequence data using Metropolis-Hastings sampling. *Genetics* 140(4): 1421-1430.

May, R.M. 1974. Biological populations with non-overlapping generations: Stable points, stable cycles and chaos. *Science* 186: 645-647.

May, R.M. 1976. Simple mathematical models with very complicated dynamics. *Nature* 262: 495-467.

McVean, G.A., S.R. Myers, S. Hunt, P. Deloukas, D.R. Bentley, and P. Donnelly. 2004. The fine-scale structure of recombination rate variation in the human genome. *Science* 304: 58104.

Meunier, J., and L. Duret. 2004. Recombination drives the evolution of GC-content in the human genome. *Mol. Biol. Evol.* 21: 984-990.

Neuhauser, C., D.A. Andow, G.E. Heimpel, G. May, R.G. Shaw, and S. Wagenius. 2003. Community genetics: Expanding the synthesis of ecology and genetics. *Ecology* 84: 545-558.

Pearl, R., and L. J. Reed. 1920. On the rate of growth of the population of the United States since 1870 and its mathematical representation. *Proc. Natl. Acad. Sci. U.S.A.* 6: 275-288.

Pritchard, J.K., M. Stephens, and P. Donnelly. 2000. Inference of population structure using multilocus genotype data. *Genetics* 155(2): 945-959.

Stephens, M. 2001. Inference under the coalescent. Pp. 213-238 in *Handbook of Statistical Genetics*. D.J. Balding, M. Bishop, and C. Cannings, eds. New York, N.Y.: John Wiley & Sons Ltd.

Strobeck, C. 1983. Estimation of the neutral mutation rate in a finite population from DNA sequence data. *Theor. Popul. Biol.* 24(2): 160-172.

Tabashnik, B.E. 1994. Evolution of resistance to *Bacillus thuringiensis*. *Annu. Rev. Entomol.* 39: 47-79.

Tajima, F. 1983. Evolutionary relationships of DNA sequences in finite populations. *Genetics* 105: 437-460.

Thompson, J.N. 1999. Specific hypotheses on the geographic mosaic of coevolution. *Am. Nat.* 153S: 1-14.

Verhulst, P.F. 1838. Notice sur la loi que la population suit dans son accroissement. *Corresp. Math. Phys.* 10: 113-121.

Wakeley, J. 2004. Recent trends in population genetics: More data! More math! Simple models? *J. Heredity* 95: 397-405.

Watterson, G.A. 1975. On the number of segregating sites in genetic models without recombination. *Theor. Popul. Biol.* 7: 256-276.

Whitham, T.G., W. Young, G.D. Martinsen, C.A. Gehring, J.A. Schweitzer, S.M. Shuster, G.M. Wimp, D.G. Fischer, J.K. Bailey, R.L. Lindroth, S. Woolbright, and C.R. Kuske. 2003. Community genetics: A consequence of the extended phenotype. *Ecology* 84: 559-573.

Wong, W.S., Z. Yang, N. Goldman, and R. Nielsen. 2004. Accuracy and power of statistical methods for detecting adaptive evolution in protein coding sequences and for identifying positively selected sites. *Genetics* 168(2): 1041-1051.

7

Understanding Communities and Ecosystems

An ecological community is an assemblage of populations of different species (plants, animals, fungi, microbes, etc.) at a given place and time. The living organisms of a community cannot be separated from their physical and chemical environment, and the combination of a community and an environment is referred to as an ecosystem. Although a community is often characterized by a dominant feature—as is, for example, a desert community or an oak savanna community—its species composition has a significant random component.

Community ecology is concerned with explaining patterns of diversity, the distribution and abundance of species within the context of these assemblages, and the underlying processes. The field of community ecology has developed rapidly over the last few decades, driven by the need to understand the consequences of anthropogenic impacts on the functioning of ecological communities.

Our understanding of how communities assemble has changed over time (Kingsland, 1991). It has ranged from regarding an ecological community as a random assemblage (Gleason, 1926) to thinking of it as a "complex organism" (Clements, 1936). In the beginning of community ecology, questions focused on community structure, population dynamics, and, in the case of plant communities, on succession (Grinnell, 1917; Clements et al., 1929). The abiotic (nonliving) environment was assigned a minor role until Lindeman's seminal paper (1942) on the trophic-dynamic aspects of ecology, which established the ecosystem as the fundamental unit.

Mathematics has played a vital role in framing community ecology concepts. Deterministic models (systems of differential or difference equations) dominated theoretical advances for much of the history of the field, and they continue to be the single most important choice of modeling framework for analytical models. In the 1920s and 1930s, two key concepts were formalized using deterministic models: competition and predation. Mathematical models greatly enhanced our understanding of both processes. Competition has been identified as an important process of ecological communities ever since Darwin proposed it as the chief mechanism in the evolution of species (Darwin, 1859). The competition models by Lotka (1932) and Volterra (1926), formulated as systems of differential equations, provide a theoretical framework for the dynamic interactions within a trophic level.[1] This framework was further developed by Elton (1927, 1933) using the concept of a niche, which he defined as "the status of an animal in its community." He linked this concept to competition in order to explain how multiple species can persist within a community. A mathematical formulation of the niche concept was finally given by Hutchinson (1957), who defined a niche as a subset of an n-dimensional hypervolume. This concept is still useful today. While the models of Lotka and Volterra describe phenomena, they lack mechanisms for competition. Tilman's (1982) resource competition model led the way from phenomenological to mechanistic competition models. Like the Lotka-Volterra models, mechanistic competition models are also based on systems of differential equations and continue to form the conceptual basis for understanding competition among multiple species.

Predation is by definition a process that occurs between trophic levels. Lotka (1925) and Volterra (1926) were the first to provide a mathematical formulation of this process, again using systems of differential equations. Differential equations model continuous time dynamics and are thus well suited for populations with overlapping generations. However, this does not always hold for biological situations. For instance, the seasonal dynamics of a host and an associated parasitoid[2] are better described by discrete time models. To include this aspect of biological realism into models, Nicholson and Bailey (Nicholson, 1933; Nicholson and Bailey, 1935) promoted systems of difference equations to describe predation models. Difference equations are now commonly employed to model interactions among species with nonoverlapping generations.

In the 1950s and 1960s the focus shifted toward understanding the

---

[1]A trophic level is a stratum of the food chain consisting of species the same number of steps from the primary source of nutrition.

[2]A parasitoid is an insect that lays its eggs in, on, or near a host and whose offspring consume the host as they develop.

relationship between the diversity of an ecological community and its stability (Real and Levin, 1991). Using qualitative arguments, Odum (1953), MacArthur (1955), and Elton (1958) concluded that diversity and stability were positively correlated. Despite the absence of carefully designed experiments and mathematical models to corroborate this claim, it remained unchallenged until May (1972, 1974) and others investigated models of randomly assembled communities whose dynamics were described by systems of differential equations, similar to those of Lotka and Volterra. These theoretical studies led to the opposite conclusion: Stability and diversity were negatively correlated. The conclusion was based on rigorous mathematics, though it lacked the synergy and validation that come from combining theoretical and empirical work. It became widely accepted by community ecologists but was questioned by ecosystem ecologists (Patten, 1975; McNaughton, 1977; Loreau et al., 2002). The diversity-stability debate was revived in the 1990s, when carefully designed experiments and mathematical models that directly addressed the variability of species abundances questioned the negative correlation between stability and diversity (see Box 7.1).

Models that try to address basic principles or processes and that focus on ideas rather than on specific biological systems have played a large role in ecology. These will be referred to as conceptual models. Many of the conceptual models of community ecology are framed as systems of differential or difference equations. This framework carries an implicit assumption of spatial homogeneity, but of course it is known that spatial movements and dispersal of individuals and spatial interactions among individuals can lead to spatial heterogeneity. Spatial movement was first included in ecological models in the 1950s with Skellam's (1951) work on the spread of muskrats. The equations were identical to those developed by Fisher (1937) to describe the spread of a novel allele. Both partial differential equations and integro-differential equations are commonly employed now to model movement and dispersal (Okubo, 1980; Holmes et al., 1994; Okubo and Levin, 2001). They are also used to investigate the effects of spatially dependent factors on the dynamics of multispecies communities. This has led to the insight that biotic interactions alone can generate spatial patterns.

Stochastic models are rarely employed in theoretical ecological studies owing to the difficulties in analyzing them, even though both environmental and demographic stochasticity play an important role in the dynamics of ecological communities. Demographic stochasticity refers to randomness that is inherent in demographic processes, such as birth or death. It is of particular importance when populations are small. Environmental stochasticity—for instance, unexplained variation in precipitation or temperature that may affect fecundity or the survival of species—can have significant effects on communities, as illustrated by the work of Chesson and Warner (Chesson and Warner, 1981; Chesson, 1994), who introduced a general modeling framework to address the role of environmental stochasticity in species coexistence. This work demonstrated the importance of nonlinear, species-specific responses to the environment that can resonate into future generations.

Demographic stochasticity has also been incorporated into individual-based spatial models where interactions among small groups of individuals are important. The study of these models was initiated by Spitzer (1970) in the United States and Dobrushin (1971) in the Soviet Union. The models are spatially explicit Markov processes, called interacting particle systems. These models were originally developed for problems in statistical physics, but it soon became clear that local interactions are important in other fields as well, including community ecology. The study of interacting particle systems and their discrete-time analogs, discrete-time cellular automata, has greatly advanced our understanding of the role of space and local interactions in the dynamics of ecological communities. This remains a very active area of research (Durrett and Levin, 1994; Neuhauser, 2001).

Interacting particle systems or cellular automata are easy to formulate, so much so that there are now numerous theoretical ecology papers that base the analysis of spatially explicit models solely on simulations. Their mathematical analysis, however, is a highly nontrivial matter. Results from simulation studies can be quite misleading, because the behavior of a finite system can differ from that of the related infinite system (Neuhauser and Pacala, 1999), and the results may not be robust with respect to the choice of local interactions (Anderson and Neuhauser, 2002). Dynamics, in particular in two spatial dimensions, may also be slow enough so that it takes a long time for the system to accurately reflect long-term behavior. For instance, the voter model (Clifford and Sudbury, 1973; Holley and Liggett, 1975) and the multitype contact process (Neuhauser, 1992; Neuhauser and Pacala, 1999) in two spatial dimensions exhibit clustering of like community members, with clusters growing indefinitely. Computer simulations have led researchers to believe that it is possible for competing species to coexist in such systems, yet rigorous mathematical analysis shows eventual exclusion of all but one type in arbitrarily large regions. This demonstrates the need for rigorous mathematical analysis.

Some analytical methods for dealing with local spatial interactions and/or stochasticity have been developed, such as metapopulation models and the moment approximation. Metapopulations are spatially implicit models (Levins, 1969; Hanski, 1999). They are formulated as systems of differential equations and track the dynamics of populations on a finite or

---

BOX 7.1
The Productivity-Stability-Diversity Debate

The relationship between productivity, stability, and diversity has been of long-standing interest, from both a purely academic point of view and a management perspective, where it has become pressing to understand the consequences of the large-scale diversity loss caused by anthropogenic disturbances. The following illustrates how increasingly more sophisticated mathematical models in combination with carefully designed experiments expand our understanding of important processes.

The past 50 years have seen a lively debate on whether diversity results in more stable and more productive ecosystems or whether the opposite is true. The arguments in favor of a positive correlation between stability and diversity in the 1950s were based on superficial comparisons between species-poor agricultural systems and species-rich tropical systems. The opposite conclusion, reached in the 1970s, was based on rigorous mathematical analysis of the equilibrium behavior of multispecies models. Early on in the discussions there was confusion, partly because different groups of researchers used different definitions of stability. The multiple definitions of stability were clarified by Pimm (1984), but the debate is still unresolved.

Loss of biodiversity can affect ecosystem processes such as nutrient cycling and energy flow. It is thus not surprising that ecosystem ecologists increasingly joined the debate on the role of biodiversity. This coming together of community ecology and ecosystem ecology since the beginning of the 1990s has helped refocus and expand the debate. A series of short-term and longer-term experiments were conducted to understand the role of biodiversity on ecosystem processes (Lawton et al., 1993; Naeem et al., 1994; Tilman and Downing, 1994). Theoretical studies soon followed.

---

infinite number of patches. Dispersal among the patches is assumed to be on a complete graph; that is, all patches are equally accessible from any other patch. Moment approximations, commonly employed in statistical physics, have proved to be useful in community ecology for studying spatial clustering (Bolker and Pacala, 1997).

The connections among the four major modeling frameworks (ordinary differential equation, partial differential equation, integro-differential equation, and interacting particle system) are well established (Durrett and Levin, 1994). As the interaction neighborhood in an interacting particle system increases either through an increase in movement relative to demographic processes or an increase in dispersal, a partial differential equation in the former case and an integro-differential equation in the latter case become good approximations; removing, in addi-

Systems of differential equations still dominated theoretical investigations, but there was an increased focus on ecosystem processes (e.g., Loreau, 1998). A different class of mathematical models found their way into the debate. Instead of deterministic systems of differential equations, where stability is based on eigenvalue properties, stochastic models were introduced that allowed keeping track of variability of both individual species and the entire community (e.g., Lehman and Tilman, 2000).

Much of the empirical and theoretical work includes only primary producers and disregards trophic links (but, see Ives et al., 2000). The potential importance of this link has been pointed out by Paine (2002). The experiments ignored belowground processes. Wardle and van der Putten (2002) point out the lack of evidence for a diversity-productivity relationship in decomposer systems. The role of symbiotic organisms also warrants further study (van der Heijden and Cornelissen, 2002). The role of biodiversity in belowground processes has only recently received attention (Freckman et al., 1997; special issue of *BioScience*, February 1999).

Theoretical work will need to be closely linked to experimental work. To guide experiments, it needs to focus on quantities that are measurable in field experiments. To have predictive power, models need to be parameterized by experimental data. Future theoretical investigations will need to include the complex interactions among different trophic levels, belowground processes, the evolutionary potential of the organisms, environmental fluctuations, and spatial structure. They will also need to address nonequilibrium behavior. There will be an increased need for long-term data in different ecosystems. The current experiments indicate that the dominant process can change over time (Fargione et al., 2004), and it will be important to provide ways to statistically test for such changes (e.g., Loreau and Hector, 2001).

tion, spatial heterogeneities results in an ordinary differential equation. Looking at this another way, if one needs to include the effects of fluctuations, correlations, and spatial heterogeneities, the simple framework of ordinary differential equations no longer suffices. Instead, the much more complicated framework of interacting particle systems (or similar processes) must be understood. The past 30 years of research in this area have considerably improved our understanding, but much work remains, because the properties of more complex multispecies assemblages embedded in ever-changing environments are only beginning to be revealed.

Analytical models will be increasingly complemented by complex simulation models that attempt to incorporate nonlinearities, nonequilibrium behavior, genetic composition, space, demographic, and environmental stochasticity. Even though (or because) computers have greatly expanded our ability to study large and complex systems, there remains a need for analytical methods. Many of the complex systems have large numbers of parameters that make exhaustive simulations nearly impossible. Developing mathematically tractable approximations of a complex simulation model can yield valuable insights into the behavior of complex models.

Ecological interactions are often complex and nonlinear and involve multiple species. Multiple stable states are a hallmark of such systems, which can lead to catastrophic changes under disturbances (Scheffer and Carpenter, 2003, and references therein). Mathematical modeling has yielded significant insights into dynamic consequences of the presence of multiple stable states. Modeling has also been applied to the recovery of systems that have undergone environmental degradation. It is often difficult to restore the original system, and it has been conjectured that this is because the system has reached a different equilibrium state (or, more generally, is in a different domain of attraction).

The importance of studying transient dynamics was pointed out by Hastings (2004). Most ecological interactions are probably far away from equilibrium. Large-scale anthropogenic perturbations, such as land-use change or nitrogen addition, are additional processes that result in nonequilibrium situations. Some of the mathematical theory has been developed, in particular when different timescales are involved. Most field experiments are studied over only short timescales, even if the dynamics are slow, thus probably describing dynamics that are not in equilibrium.

COMPUTATION

Multispecies interactions across trophic levels, including ecosystem processes, provide statistical and modeling challenges for community ecologists. The statistical analysis of large data sets that often cannot simply be analyzed using standard statistical software packages requires model development and computational methods to estimate parameters and test hypotheses. The theoretical study of large, complex systems results in models that are often analytically intractable. Computational advances have made possible the study of these models, which are currently framed as systems of differential equations. Increasingly though, a spatial component and stochastic factors are included, and both equilibrium and nonequilibrium dynamics are investigated. Few tools are currently available to deal with these frameworks when applied to large systems.

Inference in community ecology frequently deals with multiple competing hypotheses. Model selection as a way to distinguish between hypotheses provides alternatives to traditional hypothesis testing (see

Johnson and Omland, 2004, for a review). The idea here is to formulate two or more models with different embedded hypotheses, compare them with data, and analyze the goodness of fit to reveal which of the hypotheses appear to be borne out by the data. This framework was initially developed over 30 years ago (Akaike, 1973) but is only now receiving attention in ecology. It provides a way to quantify the relative support for competing hypotheses based on data. Further development of this useful tool will probably impact both experimental design and statistical analysis in ecology.

Assessment of uncertainty remains a key challenge in ecological modeling (Brewer and Gross, 2003). Few models include a stochastic component, so they are not set up to provide a distribution of results from multiple runs. In addition, different modeling approaches can yield different predictions even if the same scenarios are modeled, reflecting uncertainty in our knowledge of the underlying processes. Averaging over different models has recently been suggested as a way to increase the robustness of results (Koster et al., 2004). However, there is no general theory at this point that lends credence to such ad hoc methods.

Predictive models of ecosystems also increasingly include economic and social components. For instance, the goal of a recent National Center for Ecological Analysis and Synthesis workshop, "Global Biodiversity Scenarios" (Chapin et al., 2001), was to combine vegetation and climate models with economic and social scenarios to predict the effects of human impact on major biomes.

The management of natural ecosystems relies increasingly on sophisticated models. Spatial heterogeneity and demographic and environmental stochasticity are often key driving factors. Spatial control, a mathematically sophisticated and computationally intensive tool, appears to be a promising methodology (Hof and Bevers, 1998, 2002).

FUTURE DIRECTIONS

Interactions at the community level are influenced by and influence all other levels of organization, from genes to ecosystems, including abiotic conditions such as temperature, precipitation, and nutrient availability. For a full understanding of processes at the community level, integration across disciplines, scales, and levels of organization will be needed. The following exemplify this integration and highlight some of the mathematical developments that need to occur in order to accomplish this integration. First come the processes discussed earlier that shape ecological communities: competition and predation.

Ecology has traditionally been divided into community ecology and ecosystem ecology. Community ecology focuses on population dynamics and the interplay between the biotic and abiotic environment. Ecosystem ecology deals with fluxes of nutrients and energy. Models in community ecology describe the dynamics of biomass or individuals, whereas models in ecosystem ecology describe fluxes of matter and energy among functional units. The past 15 years have increasingly witnessed research at the interface of the two ecologies (Naeem et al., 2002). Research that addressed the diversity-stability-productivity debate illustrates this emerging synthesis of the two fields (Box 7.1). Research in this area will probably see greater integration across spatial scales and across levels of organization.

Food web studies are another example where integration across fields, scales, and levels of organization is occurring. Food webs are complex networks of interacting groups of species. A community ecology approach focuses on particular species and attempts to understand their interactions as described by competition, predation, or facilitation. A classic study by Paine (1966) illustrates this approach: Recognizing that detailed bookkeeping of the calorie consumption of the members of a food web could explain food web structure and, ultimately, the diversity of a local community, Paine manipulated food webs through removal (or addition) experiments so as to assess the importance of each link. As one of its most significant conclusions, the study demonstrated a drastic decrease in diversity after removal of the starfish *Pisaster B. glandula* from an intertidal community, thus identifying predation as an important process for maintaining diversity. Paine and Levin (1981) introduced a disturbance model that modeled the dynamics of gaps left behind by a predator and their subsequent recolonization. The model was parameterized by field data and yielded predictions that compared well with observations.

An ecosystem approach to food webs disregards species identities and instead focuses on functional groups, such as autotrophs, detritus, heterotrophs, and nutrient pools. This approach leads to compartment models that track the flux of matter and energy among the compartments. This flux is typically described by systems of differential equations.

Reiners (1986) proposed a theoretical framework for ecosystem dynamics that included both energy and nutrient considerations, calling it ecological stoichiometry. The recent book by Sterner and Elser (2002) on ecological stoichiometry provides a synthesis of processes at the cellular level to ecosystem levels based on such stoichiometry and the resulting nutrient demands of the biota. Food web models that combine both approaches are still in their early stages but have already yielded interesting insights into the importance of food quality in addition to food quantity (Loladze et al., 2000). These new models combine classical community ecology models with insights from nutrient dynamics. They are largely phenomenological but will likely become more mechanistic as our understanding of these processes across all levels of organization increases.

Additional insights into food web structure can be gained by comparing large food webs across different ecosystems. Such comparison has revealed structural commonalities, and it has been proposed that common mechanisms are responsible for network structure (Dunne et al., 2004). Recently, Brose et al. (2004) attempted to unify the relationships between species richness and spatial scaling and between species richness and trophic interactions to extend the spatial scale at which food web theory applies.

Another area of activity that requires sophisticated modeling, mathematical analysis, and statistical tools is epidemiology or, more generally, host-pathogen systems. The increased attention to disease dynamics stems from the global threat of emerging and reemerging diseases, such as avian flu, West Nile virus, or SARS. Modeling often involves much more than simple disease dynamics as embodied in the standard models of Kermack and McKendrick (1927). Human behavior, socioeconomic factors, and spatiotemporal dynamics play a significant role and must be taken into account to adequately capture the dynamics. Increasingly, researchers are studying diseases not only from a public health perspective but also with respect to how they interact with the ecological environment. Known as disease ecology, this emerging field is highly interdisciplinary, drawing from epidemiology and ecology. Complex dynamics stemming from multispecies interactions complicate the analysis and make predictions difficult. Progress in this area will require collaborations among epidemiologists, ecologists, statisticians, and mathematicians.

Microbial communities will increasingly be the focus of community and ecosystem ecology studies. They provide the opportunity for true integration across levels of organization, similar to the integration in physical systems that resulted in a description of macroscopic phenomena based on microscopic processes. Molecular biology techniques are beginning to reveal the diversity of microbes. Large-scale genome analysis is needed to assess the metabolic capacity of microbes, because proteins will need to be identified and their functions understood to reveal the metabolic pathways. Ecological studies will reveal the activity of pathways as a function of the biotic and abiotic environment; this is necessary to link the metabolic potential of microbes to community-level processes. To accomplish this integration, statistical analysis of genomic data based on evolutionary models will need to be linked to physiological models and, finally, to community-level models. Development of such models will require close collaboration between experimentalists and theoreticians. The importance of microbial studies is discussed in Box 7.2.

To illustrate the need for integration between the fields of evolution and ecology in the context of community ecology, the committee revisits a theme discussed in Chapter 6. Community ecologists largely view eco-

BOX 7.2
Microbial Ecology

Microbes are microscopic organisms that are not visible with the naked eye. They were discovered by Antony van Leeuwenhoek (1632-1723). Prokaryotic microbes (bacteria) are the oldest organisms on earth. The fossil record indicates that they evolved more than 3.8 billion years ago. Eukaryotic microbes, such as fungi and protozoa, appear to have evolved at least a billion years later.

Microbes with their unrivaled metabolic capacity play an important role in biogeochemical cycles. Since human activities have profoundly altered virtually every biogeochemical cycle, it is important to understand the roles of microbes in these cycles. Advances in molecular biology, in particular in genomics, have greatly expanded our ability to study naturally occurring microbes that have eluded us thus far owing to the difficulties in culturing them. For instance, Zehr et al. (2001) recently demonstrated that many unicellular microbes in the oxygenated region of the sea have *nif* genes, indicating that oceanic nitrogen fixation might be much higher than previously thought.

Microbes provide opportunities for integration across all levels of organization, from genes to ecosystems (Stahl and Tiedjen, 2002). Venter et al. (2004), using shotgun sequencing of microbes in the ocean, have given us a static glimpse into the enormous diversity of largely unknown organisms that are responsible for biogeochemical cycles. Their study demonstrated the feasibility of large, community-level genomics analysis to assess diversity. It is a long way from the assessment of microbial diversity to understanding the function of microbes in ecological communities. It will require integration of genomic, proteomic, and metabolomic data with community-level models. New modeling and statistical approaches will need to be developed to deal with these very large and complex systems.

System theoretical approaches are currently being championed as the key to unraveling the metabolic capacity of microbes and their role in community dynamics. An integrative approach has been suggested (Wolkenhauer et al., 2004). The complex interactions are often described by block diagrams and a network, ultimately represented through differential equations, which are the mainstay of control engineers for dealing with processes. A standard equilibrium analysis of such large systems is often not satisfying, because the systems are so complex. Modularization of networks has been suggested to understand these large complex systems (Saez-Rodriguez et al., 2004). In addition, transient dynamics might dominate much of naturally occurring communities.

It is important to realize that revealing metabolic capacity alone will not be sufficient. Environmental conditions affect the expression of metabolic pathways (Dauner and Sauer, 2001; Dauner et al., 2001). It is thus necessary to experimentally understand metabolic activities as a function of environmental conditions in order to predict community dynamics. This will require close collaboration between experimentalists and theoreticians.

logical communities as genetically homogeneous (but, see Ford, 1964). Over the last 10 years, an increasing number of studies have demonstrated the importance of including evolutionary processes in studies of ecological communities. For instance, invasive species or the assembly of novel communities can alter ecological interactions and impose strong selection on all members of a community (Reznick et al., 1997, 2001; Davis and Shaw, 2001). Evolution within a predator-prey system has been studied, for instance, by Shertzer et al. (2002) and Yoshida et al. (2003), who combined theoretical and empirical studies to demonstrate that the evolution within such a system (an algal prey and its rotifer predator) can shape population dynamics. The empirical system showed oscillations in qualitative agreement with theoretical studies. However, there was quantitative disagreement: Both the cycle period and the phase between predator and prey differed from theoretical predictions. Shertzer et al. (2002) suggested a new model that incorporated evolution of the algal prey and demonstrated that rapid evolution of the prey could explain the observed pattern. Yoshida et al. (2003) confirmed this model experimentally by growing the algal prey with and without its predator. Their study showed that resistance to the predator was a heritable trait and that there was a trade-off between resistance and competitive ability. It has been suggested that this trade-off and predation contribute to the maintenance of genetic diversity. (See Johnson and Agrawal, 2003, for a summary of these studies.) These studies demonstrate the importance of allowing genetic variation and incorporating it into ecological models. The study of these complex interactions is in its early stages. Only a combination of empirical and theoretical studies will yield much-needed insights.

The distribution and abundance of each species is a function of the whole community composition and the genetic composition of each individual in the context of the community. When ecological interactions and genetic composition of populations reciprocally affect each other, both factors need to be considered. Antonovics (1992) proposed a new framework, "community genetics" (a term suggested by J.J. Collins at Arizona State University), which is a synthesis of evolutionary genetics and community ecology and focuses on the role of genetic variation in determining community structure (Luck et al., 2003; Neuhauser et al., 2003; Whitham et al., 2003). Models that incorporate both ecological and genetic factors quickly become quite complex because they must track not only the dynamics of the species but also the genetic composition of the individuals. These models often also include a spatial component, adding to their complexity.

A community genetics perspective seems to be particularly useful when dealing with strong selection in a community context. As argued in Neuhauser et al. (2003), this is particularly likely to occur during transient dynamics following large-scale perturbations, such as habitat reduction or expansion. Habitat reduction due to land-use changes has been occurring at an unprecedented rate. The concomitant loss of genetic diversity can accelerate extinction. Habitat expansion can be observed in both agricultural and natural systems, for instance through the introduction of a novel organism such as a genetically modified organism or an exotic species invasion.

The final example illustrates the need for integration at the global scale. The effects of human activities on global climate were for the first time illustrated by Keeling et al. (1976) when they published data from Mauna Loa in Hawaii showing a clear increase in atmospheric carbon dioxide over many decades. It became clear that, in order to assess changes in the global carbon cycle, global measurements were needed. Satellite data that became available in the 1980s made it possible to estimate net primary production from remote sensing data. Satellites now capture a continuous stream of spectral data at resolutions at and below the 1-kilometer scale. For instance, the NASA Earth Observing System Terra satellite uses the Moderate Resolution Imaging Spectroradiometer (MODIS) to measure the spectral reflectance of terrestrial vegetation. This data set is used to produce a weekly data set of primary production of the entire vegetated surface, a critical quantity for assessing carbon dynamics.

Understanding carbon and nutrient cycles at global and regional scales is a very active area of ecology that integrates across community ecology and ecosystem ecology. As an example, predicting an increase in temperature as a function of an increase in carbon dioxide at the spatial scale of the whole earth was already accomplished by Arrhenius (1896). It has proved much more difficult to make predictions at regional scales, which requires linking vegetation models to global circulation models. To parameterize such models, estimates of primary production at a regional scale are needed. This will require advances in retrieving accurate estimates based on spectral information, relating those estimates to measurement of the actual state on the ground in a region, and incorporating the data into ecosystem process models. This field provides clear opportunities for linking computational models to observational data.

REFERENCES

Akaike, H. 1973. Information theory as an extension of the maximum likelihood principle. Pp. 267-281 in *Second International Symposium on Information Theory*. B.N. Petrov and F. Csaki, eds. Budapest: Akademia Kiado.

Anderson, K., and C. Neuhauser. 2002. Patterns in spatial simulations—Are they real? *Ecol. Model.* 155: 19-30.

Antonovics, J. 1992. Toward community genetics. Pp. 426-449 in *Plant Resistance to Herbivores and Pathogens: Ecology, Evolution, and Genetics*. Chicago, Ill.: University of Chicago Press.

Arrhenius, S. 1896. On the influence of carbonic acid in the air upon temperature on the ground. *The London, Edinburgh, and Dublin Philosophical Magazine and Journal of Science* 41: 237-275.

Bolker, B.M., and S.W. Pacala. 1997. Using moment equations to understand stochastically driven spatial pattern formation in ecological systems. *Theor. Popul. Biol.* 52: 179-197.

Brewer, C.A., and L.J. Gross. 2003. Training ecologists to think with uncertainty in mind. *Ecology* 84: 1412-1414.

Brose, U., A. Ostling, K. Harrison, and N.D. Martinez. 2004. Unified spatial scaling of species and their trophic interactions. *Nature* 428: 167-171.

Chapin, F.S., O.E. Sala, and E. Huber-Sannwald, eds. 2001. *Global Biodiversity in a Changing Environment: Scenarios for the 21st Century.* New York, N.Y.: Springer-Verlag.

Chesson, P. 1994. Multispecies competition in variable environments. *Theor. Popul. Biol.* 45: 227-276.

Chesson, P.L., and R.R. Warner. 1981. Environmental variability promotes coexistence in lottery competitive systems. *Am. Nat.* 117: 923-943.

Clements, F.E. 1936. Nature and structure of the climax. *J. Ecol.* 24: 252-284.

Clements, F.E., J.E. Weaver, and H.C. Hanson. 1929. *Plant Competition: An Analysis of Community Functions.* Washington, D.C.: Carnegie Institution.

Clifford, P., and A. Sudbury. 1973. A model for spatial conflict. *Biometrika* 60: 581-588.

Darwin, C. 1859. *The Origin of Species.* Reprinted in 1985. London, England: Penguin Books.

Dauner, M., and U. Sauer. 2001. Stoichiometric growth model of riboflavin-producing Bacillus subtilis. *Biotechnol. Bioeng.* 76: 132-143.

Dauner, M., T. Storni, and U. Sauer. 2001. Bacillus subtilis metabolism and energetics in carbon-limited and excess-carbon chemostat culture. *J. Bacteriol.* 183: 7308-7317.

Davis, M.B., and R.G. Shaw. 2001. Range shifts and adaptive responses to quaternary climate change. *Science* 292: 673-679.

Dobrushin, R.L. 1971. Markov processes with a large number of locally interacting components: Existence of a limit process and its ergodicity. *Problemy Peredachi Informatsii* 7: 149-164.

Dunne, J.A., R.J. Williams, and N.D. Martinez. 2004. Network structure and robustness of marine food webs. *Marine Ecol. Progress Ser.* 273: 291-302.

Durrett, R., and S.A. Levin. 1994. The importance of being discrete (and spatial). *Theor. Popul. Biol.* 46: 363-394.

Elton, C. 1927. *Animal Ecology.* London, England: Sedgwick and Jackson.

Elton, C. 1933. *The Ecology of Animals.* London, England: Methuen.

Elton, C.S. 1958. *The Ecology of Invasion by Animals and Plants.* London, England: Methuen.

Fargione, J., R. Dybzinski, C. Clark, J. Hille Ris Lambers, S. Harpole, M. Loreau, and D. Tilman. 2004. From selection to complementarity: Temporal trends in a long-term biodiversity experiment. *88th Ecological Society of America Annual Meeting*, Savannah, Georgia, August 3-8, 2003. Washington, D.C.: Ecological Society of America.

Fisher, R.A. 1937. The wave of advance of advantageous genes. *Ann. Eugen.* 7: 353-369.

Ford, E.B. 1964. *Ecological Genetics.* London, England: Methuen.

Freckman, D.W., T.H. Blackburn, L. Brussaard, P. Hutchings, M.A. Palmer, and P.V.R. Snelgrove. 1997. Linking biodiversity and ecosystem functioning of soils and sediments. *AMBIO* 26: 556-562.

Gleason, H.A. 1926. The individualistic concept of the plant association. *Bull. Torrey Botanical Club* 53: 7-26.

Grinnell, J. 1917. The niche-relationship of the Californian thrasher. *Auk* 34: 427-433.

Hanski, I. 1999. *Metapopulation Ecology.* Oxford, U.K.: Oxford University Press.

Hastings, A. 2004. Transients: The key to long-term ecological understanding? *Trends Ecol. Evol.* 19: 39-45.

Hof, J.G., and M. Bevers. 1998. *Spatial Optimization in Ecological Applications*. New York, N.Y.: Columbia University Press.

Hof, J.G., and M. Bevers. 2002. *Spatial Optimization for Managed Ecosystems*. New York, N.Y.: Columbia University Press.

Holley, R., and T.M. Liggett. 1975. Ergodic theorems for weakly interacting particle systems and the voter model. *Ann. Probab.* 3: 643-663.

Holmes, E.E., M.A. Lewis, J.E. Banks, and R.R. Veit. 1994. Partial differential equations in ecology: Spatial interactions and population dynamics. *Ecology* 75: 17-29.

Hutchinson, G.E. 1957. Population studies: Animal ecology and demography. Pp. 415-427 in *Cold Spring Harbor Symposia on Quantitative Biology*. Vol. 22. Woodbury, N.Y.: CSHL Press.

Ives, A.R., J.L. Klug, and K.Gross. 2000. Stability and species richness in complex communities. *Ecol. Lett.* 3: 399-411.

Johnson, J.B., and K.S. Omland. 2004. Model selection in ecology and evolution. *Trends Ecol. Evol.* 19: 101-108.

Johnson, M.T.J., and A.A. Agrawal. 2003. The ecological play of predator-prey dynamics in an evolutionary theatre. *Trends Ecol. Evol.* 18: 549-551.

Keeling, C.D., R.B. Bacastow, A.E. Bainbridge, C.A. Ekdahl, P.R. Guenther, L.S. Waterman, and J.F.S. Chin. 1976. Atmospheric carbon dioxide variations at Mauna Loa Observatory, Hawaii. *Tellus* 28: 538-551.

Kermack, W.O., and A.G. McKendrick. 1927. A contribution to the mathematical theory of epidemics. *Proc. Roy. Soc. London Ser. A* 115: 700-721.

Kingsland, S.E. 1991. Defining ecology as a science. Pp. 1-13 in *Foundations of Ecology: Classic Papers with Commentaries*. L.A. Real and J.H. Brown, eds. Chicago, Ill.: University of Chicago Press.

Koster, R.D., P.A. Dirmeyer, Z. Guo, G. Bonan, E. Chan, P. Cox, C.T. Gordon, S. Kanae, E. Kowalczyk, D. Lawrence, P. Liu, C.-H. Lu, S. Malyshev, B. McAvaney, K. Mitchell, D. Mocko, T. Oki, K. Oleson, A. Pitman, Y.C. Sud, C.M. Taylor, D. Verseghy, R. Vasic, Y. Xue, and T. Yamada. 2004. Regions of strong coupling between soil moisture and precipitation. *Science* 305: 1138-1140.

Lawton, J.H., S. Naeem, R.M. Woodfin, V.K. Brown, A. Gange, H.J.C. Godfray, P.A. Heads, S. Lawler, D. Magda, C.D. Thomas, L.J. Thompson, and S. Young. 1993. The Ecotron: A controlled environmental facility for the investigation of population and ecosystem processes. *Phil. Trans. Roy. Soc. Lond. B* 341: 181-194.

Lehman, C.L., and D. Tilman. 2000. Biodiversity, stability, and productivity in competitive communities. *Am. Nat.* 156: 534-552.

Levins, R. 1969. Some demographic and genetic consequences of environmental heterogeneity for biological control. *Bull. Entomol. Soc. Am.* 15: 237-240.

Lindeman, R.L. 1942. The trophic-dynamic aspects of ecology. *Ecology* 23: 399-418.

Loladze, I., Y. Kuang, and J.J. Elser. 2000. Stoichiometry in producer-grazer systems: Linking energy flow and element cycling. *Bull. Math. Biol.* 62: 1137-1162.

Loreau, M. 1998. Biodiversity and ecosystem functioning: A mechanistic model. *Proc. Nat. Acad. Sci. U.S.A.* 95: 5632-5636.

Loreau, M. 2000. Biodiversity and ecosystem functioning: Recent theoretical advances. *Oikos* 91: 3-17.

Loreau, M., A. Downing, M. Emmerson, A. Gonzales, J. Hughes, P. Inchausti, J. Joshi, J. Norberg, and O. Sala. 2002. A new look at the relationship between diversity and stability. Chapter 7 in *Biodiversity and Ecosystem Functioning: Synthesis and Perspectives*. M. Loreau, S. Naeem, and P. Inchausti, eds. Oxford, U.K.: Oxford University Press.

Loreau, M., and A. Hector. 2001. Partitioning selection and complementarity in biodiversity experiments. *Science* 412: 72-76.

Lotka, A. 1925. *Elements of Physical Biology.* Baltimore, Md.: Williams & Wilkins Co..

Lotka, A.J. 1932. The growth of mixed populations: Two species competing for a common food supply. *J. Wash. Acad. Sci.* 22: 461-469.

Luck, G.W., G.C. Daily, and P.R. Ehrlich. 2003. Population diversity and ecosystem services. *Trends Ecol. Evol.* 18: 331-336.

MacArthur, R.H. 1955. Fluctuations of animal populations and a measure of community stability. *Ecology* 36: 533-536.

May, R.M. 1972. Will large and complex systems be stable? *Nature* 238: 413-414.

May, R.M. 1974. *Stability and Complexity in Model Ecosystems.* Princeton, N.J.: Princeton University Press.

McNaughton, S.J. 1977. Diversity and stability of ecological communities: A comment on the role of empiricism in ecology. *Am. Nat.* 111: 515-525.

Naeem, S., M. Loreau, and P. Inchausti. 2002. Biodiversity and ecosystem functioning: The emergence of a synthetic ecological framework. Pp. 3-11 in *Biodiversity and Ecosystem Functioning: Synthesis and Perspectives.* M. Loreau, S. Naeem, and P. Inchausti, eds. Oxford, U.K.: Oxford University Press.

Naeem, S., L.J. Thompson, S.P. Lawler, J.H. Lawton, and R.M. Woodfin. 1994. Declining biodiversity can alter the performance of ecosystems. *Nature* 368: 734-737.

Neuhauser, C. 1992. Ergodic theorems for the multitype contact process. *Probab. Theory Rel. Fields* 91: 467-506.

Neuhauser, C. 2001. Mathematical challenges in spatial ecology. *Notices of the American Mathematical Society* 48: 1304-1314.

Neuhauser, C., D.A. Andow, G.E. Heimpel, G. May, R.G. Shaw, and S. Wagenius. 2003. Community genetics: Expanding the synthesis of ecology and genetics. *Ecology* 84: 545-558.

Neuhauser, C., and S. Pacala. 1999. An explicitly spatial version of the Lotka-Volterra model with interspecific competition. *Ann. Appl. Probab.* 9: 1226-1259.

Nicholson, A.J. 1933. The balance of animal populations. *J. Anim. Ecol.* 2: 132-178.

Nicholson, A.J., and V.A. Bailey. 1935. The balance of animal populations, Part I. *Proc. Zool. Soc. London* 3: 551-598.

Odum, E.P. 1953. *Fundamentals of Ecology.* Philadelphia, Pa.: Saunders.

Okubo, A. 1980. *Diffusion and Ecological Problems: Mathematical Models.* Biomathematics, Vol. 10. New York, N.Y.: Springer.

Okubo, A., and S.A. Levin. 2001. *Diffusion and Ecological Problems: Modern Perspectives.* New York, N.Y.: Springer.

Paine, R.T. 1966. Food web complexity and species diversity. *Am. Nat.* 100: 65-75.

Paine, R.T. 2002. Trophic control of production in a rocky intertidal community. *Science* 296: 736-739.

Paine, R.T. , and S.A. Levin. 1981. Intertidal landscapes: Distribution and dynamics of pattern. *Ecol. Monogr.* 51: 145-178.

Patten, B.C. 1975. Ecosystem linearization: An evolutionary design problem. *Am. Nat.* 109: 529-539.

Pimm, S.L. 1984. The complexity and stability of ecosystems. *Nature* 307: 321-326.

Real, L.A., and S.A. Levin. 1991. The role of theory in the rise of modern ecology. Pp. 177-191 in *Foundations of Ecology: Classic Papers with Commentaries.* L.A. Real and J.H. Brown, eds. Chicago, Ill.: University of Chicago Press.

Reiners, W.A. 1986. Complementary models for ecosystems. *Am. Nat.* 127: 59-73.

Reznick, D.N., M.J. Butler, and F.H. Rodd. 2001. Life-history evolution in guppies. VII. The comparative ecology of high- and low-predation environments. *Am. Nat.* 157: 126-140.

Reznick, D.N., F.H. Shaw, F.H. Rodd, and R.G. Shaw. 1997. Evaluation of the rate of evolution in natural populations of guppies (*Poecilia reiculata). Science* 275: 1934-1937.

Saez-Rodriguez, J., A. Kremling, H. Conzelmann, K. Bettenbrock, and E.D. Gilles. 2004. Modular analysis of signal transduction networks. *IEEE Control Systems Magazine* 24(4): 35-52.

Scheffer, M., and S.R. Carpenter. 2003. Catastrophic regime shifts in ecosystems: Linking theory to observation. *Trends Ecol. Evol.* 18: 648-656.

Shertzer, K.W., S.P. Ellner, G.F. Fussmann, and N.G. Hairston Jr. 2002. Predator-prey cycles in an aquatic microcosm: Testing hypotheses of mechanism. *J. Anim. Ecol.* 71: 802-815.

Skellam, J.G. 1951. Random dispersal in theoretical populations. *Biometrika* 38: 196-218.

Spitzer, F. 1970. Interaction of Markov processes. *Adv. Math.* 5: 246-290.

Stahl, D.A., and J.M. Tiedje. 2002. *Microbial Ecology and Genomics: A Crossroads of Opportunity.* American Academy of Microbiology Critical Issues Colloquia Report. Washington, D.C.: American Society for Microbiology.

Sterner, R.W., and J.J. Elser. 2002. *Ecological Stoichiometr.* Princeton, N.J.: Princeton University Press.

Tilman, D. 1982. *Resource Competition and Community Structure.* Princeton, N.J.: Princeton University Press.

Tilman, D., and J.A. Downing. 1994. Biodiversity and stability in grasslands. *Nature* 367: 363-365.

van der Heijden, M.G.A., and J.H.C. Cornelissen. 2002. The critical role of plant-microbe interactions on biodiversity and ecosystem functioning: Arbuscular mycorrhizal associations as an example. Pp. 181-194 in *Biodiversity and Ecosystem Functioning: Synthesis and Perspectives.* M. Loreau, S. Naeem, and P. Inchausti, eds. Oxford, U.K.: Oxford University Press.

Venter, J.C., K. Remington, J.F. Heidelberg, A.L. Halpern, D. Rusch, J.A. Eisen, D. Wu, I. Paulsen, K.E. Nelson, W. Nelson, D.E. Fouts, S. Levy, A.H. Knap, M.W. Lomas, K. Nealson, O. White, J. Peterson, J. Hoffman, R. Parsons, H. Baden-Tillson, C. Pfannkoch, Y.H. Rogers, and H.O. Smith. 2004. Environmental genome shotgun sequencing of the Sargasso Sea. *Science* 304: 66-74.

Volterra, V. 1926. Variazioni e fluttuazioni del numero d'individui in specie animali conviventi. *Mem. R. Accad. Naz. dei Lincei. Ser. VI* 2: 31-113.

Wardle, D.A., and W.H. van der Putten. 2002. Biodiversity, ecosystem functioning and above-ground-below-ground linkages. Pp. 155-168 in *Biodiversity and Ecosystem Functioning, Synthesis and Perspectives,* M. Loreau, S. Naeem, and P. Inchausti, eds. New York, N.Y.: Oxford University Press.

Whitham, T.G., W.P. Young, G.D. Martinsen, C.A. Gehring, J.A. Schweitzer, S.M. Shuster, G.M. Wimp, D.G. Fischer, J.K. Bailey, R.L. Lindroth, S. Woolbright, and C.R. Kuske. 2003. Community and ecosystem genetics: A consequence of the extended phenotype. *Ecology* 84: 559-573.

Wolkenhauer, O., B.K. Ghosh, and K.-H. Cho. 2004. Control and coordination in biochemical networks. *IEEE Control Systems Magazine* 24(4): 30-34.

Yoshida, T., L.E. Jones, S.P. Ellner, G.F. Fussmann, and N.G. Hairston Jr. 2003. Rapid evolution drives ecological dynamics in a predator-prey system. *Nature* 424: 303-306.

Zehr, J.P., J.B. Waterbury, P.J. Turner, J.P. Montoya, E. Omoregie, G.F. Steward, A. Hansen, and D.M. Karl. 2001. New nitrogen-fixing unicellular cyanobacteria discovered in the North Pacific subtropical gyre. *Nature* 412: 635-638.

8

Crosscutting Themes

The organization of this report around levels of biological organization reflects the committee's view that the interplay between mathematics and biology during the 21st century will be driven by biological problems. Nonetheless, the committee also recognized that this view of the mathematics-biology interface risks the neglect of crosscutting themes—that is, mathematical ideas or areas of productive research activity that cut across levels of biological organization, emerging and re-emerging in diverse biological contexts. Accordingly, it concludes its report with a few examples of such themes, starting with some mathematical ideas that have assumed central importance at the interface between mathematics and biology.

THE "SMALL n, LARGE P" PROBLEM

Classical statistics largely arose in settings where typical problems involved estimating a small set of parameters (P) from large numbers of data points (n). Modern examples of such "small P, large n" problems include estimating the overall inflation rate—or, perhaps, a modest number of category-specific inflation rates—from longitudinal data on the prices of large numbers of specific items. Of course, similar problems often also arise in overtly biological contexts (e.g., analysis of life expectancies or the dose-response characteristics of pharmaceutical agents); however, these applications of classical statistics to living systems are typically far removed from considerations of underlying biological mechanism. In small P, large n contexts, estimates of the parameter set are expected to improve as the number of data points increases, and much of the machinery of classical statistics addresses trade-offs between the reliability of parameter estimates and the number of samples analyzed.

In many biological research settings, the statistical challenge is quite different. Individual experiments—for example, a microarray-based measurement of the levels of thousands of messenger-RNA levels in a single RNA sample extracted from a particular tumor—are often information-rich. However, the number of independent measurements from which a biologist seeks to draw conclusions (e.g., the number of tumors analyzed) may be quite small. Similarly, geneticists are now contemplating measuring the genetic variants present at $\sim 10^5$ sites across the genomes of individual research subjects even though the number of individuals—for example, the number of cases and controls in a disease-susceptibility study—is under severe practical constraints. The challenge in these situations is analogous to attempting to reach conclusions from a moderate number of photographs of, for example, profitable and unprofitable restaurants. Although increasing the number of restaurants photographed would certainly improve the reliability of the study, presuming that the sampling strategy was well considered, success would depend even more heavily on strategies for representing and modeling the immense amount of information in each photograph. Interest among biologists in problems with similar statistical properties has grown dramatically in recent years. The committee examines this phenomenon here as a prime example of a crosscutting mathematical theme on the interface between mathematics and contemporary biology. It illustrates both the progress that has been made during the past decade and the challenges that lie ahead.

Finding Patterns in Gene-Expression Data

Although the small n, large P problem is encountered in many biological contexts, the challenges of interpreting gene-expression data provide a prototypical example that is of substantial current interest. The development of microarray technology, which can yield the transcription profiles of $>10^4$ genes in a single experiment, has enabled global approaches to understanding regulatory processes in normal or disease states. Substantial work has been done on the selection and analysis of differentially expressed genes for purposes ranging from the discovery of new gene functions and the classification of cell types to the prediction of clinically important biological phenotypes (*Nature Genetics* Supplement 21, 1999; Golub et al., 1999; Tamayo et al., 1999; *Nature Genetics* Supplement 32, 2002). The best of the markers that have emerged from this research have already shown promise for both diagnostic and prognostic clinical use.

Applications to tumor classification have attracted particular interest since it has been estimated that over 40,000 cancer cases per year in the United States present major classification challenges for existing clinical and histological approaches. Gene-expression microarrays for the first time offer the possibility of basing diagnoses on the global-gene-expression profile of the tumor cells. Moreover, the discovery of gene-expression patterns that are significantly correlated with tumor phenotype can clarify molecular mechanisms of pathogenesis and potentially identify new strategies for treatment (Shipp et al., 2002). Similar opportunities exist for many other poorly understood diseases. For example, the recent discovery that a set of genes in the oxidative-phosphorylation pathway is more highly expressed in the muscle biopsies of normal controls than in those of patients with Type 2 diabetes has opened new avenues in diabetes research (Mootha et al., 2003). Closer study of the most highly correlated genes in this set led to the hypothesis that PGC-1α might regulate this subset of genes, a result that was then confirmed by further laboratory study. By this route, an aberration in PGC-1α expression has become a prime candidate for being a step in disease progression.

The pattern-recognition techniques required for analyzing gene-expression data and other large biological data sets are often called supervised and unsupervised learning. Machine-learning tools based on these techniques, designed in collaborations between bioscientists and mathematical scientists, have already come into widespread use. Pattern recognition via supervised and unsupervised learning is based on quantitative, stochastic descriptions of the data, sometimes referred to as associative models. These models typically incorporate few or no assumptions about the mechanistic basis for the patterns that they seek to discover.

In unsupervised learning techniques, the structure in a data set is elucidated without using any a priori labeling of the data. Unsupervised learning can be useful during exploratory analysis. Supervised techniques create models for classifying data by training on labeled members of the classes that are to be distinguished—for example, invasive and noninvasive tumors. Supervised techniques have an advantage over unsupervised techniques because they are less subject to structure that is not directly relevant to the distinction of interest, such as the laboratory in which the data were collected. Unfortunately, training sets are not available in many biological situations.

General machine-learning algorithms that are potentially useful in this area of research stem from fields such as psychology and systematics. The algorithms have then been refined in such fields as financial modeling or market analysis, where the number of independent instances— that is, data points such as days of observation or individual-customer transactions—is substantially larger than the number of variables measured (i.e., the dimension of the problem; see Hastie et al., 2001). A rule of thumb is that the number of data points should be at least as large as the square of the number of dimensions (Friedman, 1994). This goal is often out of reach in biological-pattern-recognition studies: Typically, the data sets available comprise a small number of samples per biological class (generally fewer than 100, often only 10 or 20); however, this small number of samples contrasts with the large number of features that characterize each sample—for example, expression levels on the order of 10^4 genes). In many situations, increasing the number of samples is simply not possible. Hence, biological applications of machine learning often involve instances of the small n, large P problem discussed above.

The first generation of discovery and recognition tools suitable for the analysis of microarray data has been built, and these tools have established that expression data can be productively mined for purposes such as tumor classification (Golub et al., 1999; Bittner et al., 2000; Slonim et al., 2000; Tamayo et al., 1999; Perou et al., 1999; Perou et al., 2000), chemosensitivity of tumors (Staunton et al., 2001), and treatment outcome (Alizadeh et al., 2000). The classifiers used in these papers are still heavily employed today and are being refined to apply to cases where subtle signatures of phenotypes such as post-treatment outcome are the endpoints. Despite clear successes in applying machine learning to gene-expression data, most studies have oversimplified the problem by treating genes as independent variables. Even when coregulation is taken into account (Cho et al., 1998; Eisen et al., 1998), existing methods still fail to capture the complex patterns of interaction that characterize all biological regulatory processes. They also address inadequately the diversity of biological mechanisms that can lead to indistinguishable phenotypes.

In the context of microarray data, although measurements may define a very high-dimensional space of >10,000 genes, the expression levels of these genes are dependent variables. In typical cases, a smaller set of variables—on the order of a few hundred metagenes—adequately captures the process being modeled. Thus, the problem can be approached by reducing the gene dimension in a principled way to reach the desired small P, moderate n level with which traditional statistical theory generally deals. However, major challenges remain in learning how to use a reduction process that actually reflects the biological mechanisms that lead to the highly correlated expression of the members of particular subsets of genes. These features of the problem highlight the need to develop statistical frameworks that will accommodate both the presence of many irrelevant variables and the high interdependence of those variables that are relevant. Next, the committee presents a brief description of the existing statistical framework.

Supervised Learning

In the supervised-learning setting, there exists a data set of samples that belong to two different phenotypes, Class A and Class B. The goal is to build a model that when presented with a new sample of unknown phenotype can identify its corresponding class with high accuracy. Mathematically, the challenge is to infer a function F that assigns a phenotype label A or B to a point $G = (g_1, g_2, \ldots, g_n)$ in a high-dimensional space—the expression levels of ~10,000 genes in each sample—from a small number of $(G, F(G))$ pairs. In general, such systems are highly underdetermined. Moreover, as discussed above, the variables are not independent: Genes are expressed in response to the activation of biochemical pathways resulting from multiple gene products. While this interdependence of the variables is the only reason the problem is tractable, it severely limits the utility of traditional statistical approaches for testing the significance of observations.

Classifiers tend to overfit the data—that is, they have poor predictive power outside the training sample because of the small number of samples, the large number of features (high dimensionality), and noise in the data. To address this problem, the number of variables must be reduced by selecting a subset of features that are most highly correlated with the phenotype distinction. From the perspective of machine learning and pattern recognition, the problem of optimal feature selection is intractable, and biologists must be content with empirical approximations that are tailored to the specific application (Duda et al., 2000). Traditional methods for determining the statistical significance of the features—for example, the expression levels of particular genes—to be used as classifiers assume a known underlying distribution of values and independence of the features. However, a good parametric description for expression values has yet to be determined and may not exist. Gene-gene interactions are fundamental to biological processes, and thus gene-expression data are inherently incompatible with independence assumptions.

Some groups have used permutation-based methods to solve the gene-selection problem (Golub et al., 1999; Slonim et al., 2000; Tusher et al., 2001). In these methods, one compares the observed distribution of gene correlations with phenotype against a distribution obtained by randomly assigning class labels to samples. Permuting the class labels preserves the gene-gene dependencies within the data set. Other methods include using step-down adjusted-p values (Dudoit et al., 2002), generalized likelihood tests (Ideker et al., 2000), Bayes hierarchical models (New- ton et al., 2001; Baldi and Long, 2001), and combined data from replicates to estimate posterior probabilities (Lee et al., 2000). So far, no systematic comparisons of the error rates and statistical power of these different methods have been published. Clearly this is an area that needs more research and a strong formal framework.

Even the question of how many samples might be needed to improve the accuracy of the original classifier or to provide a more rigorous statistical validation of the predictive power of classifiers is difficult. Traditional power calculations (Adcock, 1997) do not address the situation posed by gene-expression data: They estimate the confidence of an empirical error estimate based on a given data set, not how the error rate might decrease given more data. Attempts have been made to answer the latter question using nonparametric methods and permutation testing (Cortes et al., 1993; Cortes et al., 1995; Mukherjee et al., 2003), but formal analysis of this problem remains an open challenge.

One widely used approach to supervised learning involves the use of support vector machines (SVMs). SVMs are based on a variation of regularization techniques for regression (Vapnik, 1998; Evgeniou et al., 2000) and are related to a much older algorithm, the perceptron (Minsky and Papert, 1988; Rosenblatt, 1962). Perceptrons seek a hyperplane to separate positive and negative examples. The SVM seeks to further separate such examples. It is trained by solving a convex optimization problem, usually involving a large number of variables. The objective function involves a penalty, which has to be tuned to avoid overfitting. Performance of the SVM is reasonably measured by the proportion of misclassified cases in a new sample. Since such a sample is not usually available, methods involving splitting the original training sample—known as cross validation—are used. Other promising methods such as "boosting" are being developed in the statistics and machine-learning communities (Hastie et al., 2001). A basic limitation on all such methods is that, even when they identify reproducibly observable clusters, they may not provide insight into the biological mechanisms that underlie the process or phenotype being studied.

Unsupervised Learning

In unsupervised learning, the data are not labeled. The goal is to determine the underlying structure of a data set and to uncover relevant patterns and possible subtypes that can then provide the starting point for additional biological characterization. Many types of clustering algorithms have been applied to expression data—for example, hierarchical clustering (Cho et al., 1998; Eisen et al., 1998), self-organizing maps (SOM)

(Tamayo et al., 1999), and k-means. These methods focus on the dominant structure present in a data set while potentially missing more subtle patterns that might be of equal or greater biological interest.

In contrast, there are a number of local, or bottom-up, unsupervised methods that seek to identify and analyze subpatterns in gene expression data: the SPLASH algorithm (Califano, 2000), conserved X motifs (Murali and Kasif, 2003), the PLAID algorithm (Lazzeroni and Owen, 2002), the association rules of Becquet et al. (2002), or the frequent itemsets and modules of Tamayo et al. (2004) and Segal et al. (2004). Bottom-up approaches provide a comprehensive catalog of subpatterns and expose most or all of the potentially interesting structure. They tackle the small n, large P problem by attempting to directly extract and isolate the relevant signals. The challenge is the difficulty of dealing with the potentially large number of patterns discovered by these methods, many of which are typically false positives. The small n, large P problem remains in trying to find appropriate filters to separate real patterns from noise and finding ways to assemble the discovered patterns into a coherent representation of the data. Unfortunately, there is no theoretical foundation for evaluating the significance of extracted subpatterns purely on the basis of the data.

Classical approaches to reduce the noise and dimensionality use global decompositions or projections of the data that preserve the dominant structure. Examples of these methods include principal-component analysis (PCA) (Bittner et al., 2000; Pomeroy et al., 2002), singular-value decomposition (SVD) (Alter et al., 2000; Kluger et al., 2003) and PLAID (Lazzeroni and Owen, 2002). Unsupervised global, or top-down, approaches address the small n, large P problem by using appropriate projections from gene space to find a set of molecular coordinates that captures dominant signals. Once again, these methods often produce difficult-to-interpret, complex, or unwieldy representations of the data.

Projection algorithms such as nonnegative matrix factorization (NMF) (Lee and Seung, 1999; Kim and Tidor, 2003; Brunet et al., 2004) represent a new generation of methods that attempt to project the data into the space of a small number of metagenes, which provide representations that aid in biological interpretation and have the potential to guide follow-up experiments in the laboratory. NMF is based on a decomposition-by-parts approach, which was introduced by Lee and Seung (1999) to identify characteristic features of faces and semantic features of text. Despite its usefulness and practical success in clustering data, there are many open questions concerning the algorithm, its convergence properties, and the properties of the projected representation. Recent research on supervised-learning problems focuses on low-dimensional, nonlinear-manifold representations (Roweis and Saul, 2000; Tenenbaum et al., 2000), or other nonlinear sparse representations. This work is in its infancy and has not yet been systematically used for biological applications.

All the supervised and unsupervised approaches the committee describes here have associated questions that require further investigation. Some of these questions follow: Is it possible to develop a formal framework for evaluating the significance of features or subpatterns extracted in a small n, large P context? What is the best way to determine the correct number of clusters within a given data set? How does one validate clustering or decomposition results? How does one compare the correctness of two decompositions of a data set? None of these challenges is unique to biology, but biological applications bring them to the fore.

ANALYSIS OF ORDERED SYSTEMS

Systems or processes with strong spatial or temporal order are ubiquitous in biology. Examples involve the sequence of bases in the genome, the propagation of nerve impulses, and—at a higher level of biological organization—animal behavior. Mathematical techniques for analyzing ordered processes have been successfully imported into biology from other research areas. A particularly important example is the hidden Markov model (HMM). HMMs have been used in areas such as speech recognition since the 1970s. More recently, they have been applied with great success in many areas of biology. HMMs require more specific modeling of the structure within a data set than do the nonparametric methods discussed in the preceding section. When suitable models exist, this requirement is a strength: Indeed, it is sometimes possible to make valid inferences from a single instance of a biological entity such as a gene— that is, to analyze a small n, large P problem when $n = 1$. This escape from the small n, large P problem is somewhat illusory since the HMM assumption enables us to use the large number of bases in the single gene to provide us with nearly identically distributed and independent proxy samples.

Applications of Hidden Markov Models to the Analysis of DNA, RNA, and Protein Sequences

An HMM describes a set of states connected by transitions between states. The transitions occur according to a Markov process. This means that the distribution of the mth state in the series, given the preceding $m - 1$ states, depends only on the $(m - 1)$st state. However, the states themselves are not observed (they are hidden): They reveal themselves by emitting observable variables. In speech recognition, the observed variables might be phonemes. In DNA and protein applications, they would be the nucleotides or amino acids corresponding to specific sequences. All of the parameters of the HMMs governing the emissions of variables from specific states and the transitions between states are probabilities. There are many well-established algorithms for addressing important questions that arise during the use of HMMs. For example, given an HMM and a sequence, one can determine the probability that the sequence was generated by the HMM. Calculation of these probabilities allows one to find within a set of candidate models the HMM that is most likely to have generated a particular sequence. One can also find the specific path of the sequence through the HMM. This capability allows one to parse the sequence into the most likely arrangement of hidden states. Note, however, that these are still associative rather than mechanistic models and are usually viewed simply as very crude approximations to reality. Two specific applications of HMMs to biological sequences, profile HMMs for protein families and HMMs for predicting gene structures in DNA, are discussed below.

Profile HMMs

Profiles for protein families were introduced by Gribskov et al. (1987) as a method for representing the variability in protein sequences of the same family. Given an alignment of the sequences, the profile provides a score for all possible amino acids that might occur at each position and also a score associated with deletions and insertions at different positions. Profile HMMs were introduced by Krogh et al. (1994) to put the concept of a profile in a fully probabilistic framework. The hidden states, which are the positions in the protein-family model, are hidden because any individual sequence may have insertions and deletions relative to the model. Given a set of sequences known to be of the same protein family, expectation maximization (EM) can be used to determine the parameters of the HMM for the family. Given a profile HMM for a specific family and a protein sequence, one can determine the best alignment of the sequence to the family and the probability that the protein sequence would be generated by the HMM for the family. One can then classify proteins into different families by comparing those probabilities.

The emission probabilities at each position of the HMM can indicate important features of a protein family. For example, active-site residues in enzymes tend to be highly, if not completely, conserved among all members of a family. Positions that are all hydrophobic are likely to be in the interior of the protein or exposed to hydrophobic environments such as the interior of membranes. Given a set of HMMs for different protein families and at least one known structure for each family, HMM-based methods provide an effective means for predicting the approximate structure of a new protein from its sequence simply by determining the family to which the protein is most likely to belong. Of course, if the protein does not belong to any of the established families, this approach fails, and one must resort to ab initio methods. However, as increasing numbers of protein structures are determined and it becomes increasingly clear that most proteins—or at least domains of proteins—fall into a limited set of structural classes, HMM-based classification methods are providing more and more useful predictions of protein structure and function.

Despite past success, there is ample room for improvement in the development and application of HMMs to protein families. Two important areas for improvement deal with nonindependence in the data. Usually it is assumed that the protein sequences from which a profile HMM is built are independent samples from the set of sequences in the family. In actuality, members of the sample set are related to each other by a phylogenetic tree, and means of incorporating that information into Profile HMMs should improve their performance. The other nonindependence issue involves limitations on the structures of the HMMs themselves. Profile HMMs assume that the positions are independent of one another or, at most, that there is a low-order Markov dependence among nearby positions. In reality, distant positions within the protein may be interacting with one another, and the amino acid frequencies at these interacting sites may be correlated. Such long-distance correlations occur frequently in RNA structures and are represented by higher-order models called stochastic context-free grammars. However, even stochastic context-free grammars are limited to correlated positions that are nested. This condition does not hold for typical protein interactions; indeed, it does not even apply to all intramolecular interactions within RNA molecules. Finding efficient ways of taking such long-range interactions into account, while maintaining the advantages of probabilistic models, would provide an important improvement, especially for structure prediction.

HMMs in Gene Finding

Gary Churchill (1989) first applied HMMs to partition DNA sequences into domains with different characteristics. Early on, David Searls (1992) recognized the analogy between the parsing of sequences in linguistic analysis and the determination of functional domains in DNA sequences. By the early 1990s, David Haussler and colleagues had begun applying HMMs to the problem of identifying the protein-coding regions in genomic DNA sequences (see Krogh et al., 1994, Stormo and Haussler, 1994; Kulp et al., 1996). By that time, large-scale DNA-sequencing projects had begun, and there were many DNA sequences in the databases with no known associated genes or functions. Predicting what proteins might be encoded in these newly discovered DNA sequences was an important problem.

The basic structure of an HMM maps well to the gene-prediction problem. The hidden states are the functional domains of the DNA sequence: For example, some regions of the DNA code for protein sequence, other regions code for untranslated portions of genes, while still others are intergenic. Each class of regions has some statistical features that help to distinguish it from the other classes. For example, protein-coding exons must have an open reading frame and often use codons in a biased manner, so the base-emission probabilities characterizing that state will be different from those characterizing introns or other classes. There is also a clearly defined grammar for protein-coding regions: Introns must alternate with exons, and intergenic regions must surround these alternating exon-intron segments.

On the other hand, some aspects of gene structure are not captured by simple HMM architectures. For example, when introns are removed, the two joined exons must remain in-frame, so the HMM has to maintain a memory of the reading frame from the previous exon as it passes over the intron. Furthermore, exons and introns have different length distributions; neither is simply geometric, as would be modeled by a simple HMM. Finally, the boundaries between domains are often indicated by signals in the DNA sequence—that is, specific sequence motifs that are themselves modeled by the probability distributions of bases at different positions within the motifs.

Gene-prediction accuracy can be improved by incorporating other evidence that is not derived from the DNA sequence alone—for example, similarities between the protein sequence inferred from the predicted gene structure and previously known protein sequences. To utilize all the different kinds of information that are useful for gene prediction and to capture the details of gene structures, HMMs have been extended to generalized HMMs (GHMMs) (Kulp et al., 1996; Burge and Karlin, 1997). These new models, which couple classical HMMs to machine-learning techniques, provide significantly better predictions than previous models. Recently, the methodology was extended to predict simultaneously gene structure in two homologous sequences (Korf et al., 2001; Meyer and Durbin, 2002; Alexandersson et al., 2003). Since corresponding (orthologous) genes in closely related organisms are expected to have similar structures, adding the constraint that the predicted structure be compatible with both sequences can significantly improve accuracy.

Despite these advances, there is still much room for improvement in gene prediction. Overall accuracy, even when using two species, is far from 100 percent. Increasingly, the failures of gene-prediction methods are due to the inherent biological complexity of the problem. Recent data have emphasized that a region of DNA may code for multiple protein variants owing to alternative splicing. Indeed, it now appears that the majority of human genes are alternatively spliced to give two or more protein products. This biological reality means that the basic assumption of gene-prediction HMMs—that any particular base in the sequence derives from a unique hidden state rather than playing multiple functional roles—is incorrect. It may be possible to extend HMMs to deal with such situations by making explicit states that accommodate dual roles or by predicting alternative products from the optimal and suboptimal predictions of the HMMs.

Much remains to be learned about the various classes of DNA segments and the features that define them. In particular, regulatory regions pose major challenges. These regions are composed of sets of binding sites for regulatory proteins, organized into modules that control gene expression. More experimental information is needed to incorporate the properties of regulatory regions into gene-prediction models. However, eventually it may be possible not only to predict what proteins are encoded by a given DNA region but also to predict the conditions under which they are expressed.

APPLICATIONS OF MONTE CARLO METHODS IN COMPUTATIONAL BIOLOGY

The early development of dynamic Monte Carlo methods (Metropolis et al., 1953) was motivated by the study of liquids and other complex physical systems. Increasing computational power and theoretical advances subsequently expanded their application throughout many areas of science, technology, and statistics. The use of dynamic Monte Carlo methods in statistics began in the early 1980s, when Geman and Geman (1984) and others introduced them in the context of image analysis. It was quickly realized that these methods were also useful in more traditional applications of parametric statistical inference. Tanner and Wong (1987), as well as Gelfand and Smith (1990), pointed out that such standard statistical problems as latent-class models, hierarchical-linear models, and censored-data regression all have structures allowing the effective use of iterative sampling when estimating posterior and predictive distributions. Within the past decade, there has been an explosion of interest in the application of Monte Carlo methods to diverse statistical problems such as clustering, longitudinal studies, density estimation, model selection, and the analysis of graphical systems (for reviews, see Tanner (1996); Gilks et al. (1996); Liu (2001)). Concomitant with the spread of Monte Carlo methods in model-based analysis, there has been a general increase in reliance on computational inference in many areas of science and engineering.

Computational inference based on Bayesian or likelihood models often leads to large-scale Monte Carlo sampling as a global optimization strategy. In summary, in areas extending far beyond biology, Monte Carlo sampling has become an important tool in scientific computation, particularly when computational inference is based on statistical models. The committee describes below some uses of Monte Carlo methods in computational biology and discusses the limitations on current methods and possible directions for future research.

Gibbs Sampling in Motif Finding

The identification of binding sites for transcription factors that regulate when and where a gene may be transcribed is a central problem in molecular biology. Beginning in the late 1980s, this problem was formulated as a statistical-inference problem by Gary Stormo, Charles Lawrence, and others. It was assumed that the upstream regions of a set of coregulated genes are enriched in binding sites that have nucleotide frequencies different from the background sequences. In general, neither the site-specific nucleotide frequencies (the motif model) nor the locations of the sites are known. Currently, the most successful algorithm for the simultaneous statistical inference of the motif model and the sites involves application of a version of the Monte Carlo algorithm called the Gibbs sampler (Lawrence et al., 1993). Computational biologists are presently working to extend this basic approach to incorporate cooperative interactions between bound transcription factors and to analyze sequences from multiple species that are evolutionarily related.

Inference of Regulatory Networks

Probabilistic networks were developed independently in statistics (Lauritzen and Spiegelhalter, 1988) and computer science (Pearl, 1988). Directed-graph versions of probabilistic networks, known as Bayesian networks, have played an important role in the formulation of expert systems. Recently, Bayesian networks also proved to be useful as models of biological regulatory networks (Friedman et al., 2000). In these networks, the genes and proteins in a regulatory network are modeled as nodes in a directed graph, in which the directed edges indicate potential causal interactions—for example, gene A activates gene B. Given the network structure—that is, the graph structure specifying the set of directed edges—there are efficient algorithms for inferring the remaining parameters of the network. If the network structure is unknown, inferring it involves sampling from its posterior distribution, given the data. This computation is challenging, since the space of all possible network structures is superexponentially large. The development of Monte Carlo schemes capable of handling this computation would be of great value in computational biology.

Sampling Protein Conformations

The protein-folding problem has been a grand challenge for computational molecular bioscientists for more than 30 years, since Anfinsen demonstrated that the sequences for some proteins determine their folded conformations (Sela et al., 1957). To formulate the computational problem, one sets up an energy function based on considerations of bonding geometry, as well as electrostatic and van der Waals forces. Possible conformations of the protein (i.e., the relative spatial positions of all its heavy atoms) can then be sampled either by integrating Newton's second law (i.e., carrying out a molecular dynamic calculation) or by Monte Carlo sampling of the corresponding Boltzmann distribution (for a review of this, see Frenkel and Smit, 1996). This problem is attractive both because it is intrinsically important for understanding proteins and because computational results can be compared with experimentally solved structures. Hence, unlike in many other areas of predictive modeling in biology, there are easily applied, objective criteria for comparing the relative accuracy of alternative models. At present, de novo computation of native protein structures is not feasible. Thus, the near-term focus of most research in this area is on gaining an improved understanding of the mechanism of protein folding (Hansmann et al., 1997; Hao and Scheraga, 1998). Monte Carlo methods are important in these investigations because they provide wider sampling of the conformation space than do conventional methods. The study of folding-energy landscapes is generally based on a simplified energy function—for example, effects of entropy in the solvent are incorporated into artificial hydrophobic terms in the energy function—and a greatly simplified conformation space. Even with such simplifications, Monte Carlo methods are often the only way to sample this space.

LESSONS FROM MATHEMATICAL THEMES OF CURRENT IMPORT

This discussion of flourishing applications of machine learning, hidden Markov models, and Monte Carlo sampling illustrates how particular mathematical themes can gain prominence in response to trends in biological research. The advent of high-throughput DNA sequencing and gene-expression microarrays brought to the forefront of biological research large amounts of data and many classes of problems that demanded the importation of broad, powerful mathematical formalisms. Continued reliance on ad hoc solutions to particular problems would have impeded the development of whole areas of biology. In the instances discussed, the biological problems that needed solution were sufficiently analogous to problems previously encountered in other fields that relevant mathematical formalisms were available. As these formalisms came into widespread use in the biosciences, particular limitations, associated in many instances with the general characteristics of the biological problems to which they were applied, became evident and stimulated new mathematical research on the methods themselves.

The committee expects this dynamic to recur as mathematical biology matures. Indeed, the committee attached more importance to the process than to its particular manifestations in the 1990s and early 2000s. While the techniques described here have broad importance at the moment, the committee does not expect them to dominate the biosciences over the long term. Indeed, as it did in the Executive Summary and Chapter 1, "The Nature of the Field," the committee once more cautions against drawing up a list of mathematical challenges that are not grounded in specific biological problems. Both the biosciences and mathematics have strange ways of surprising us. Mathematics can be useful in ways that are not predictable. For example, Art Winfree's use of topology provided wonderful insights into the way many oscillatory biological processes work (Winfree, 1983). Similarly, De Witt Sumners's use of topology to understand aspects of circular DNA (Sumners, 1995) and Gary Odell's topological observations about the gene network behind segment polarity were quite unexpected (von Dassow et al., 2000). Yet, even though topological arguments have provided biologists with powerful insights, the committee did not conclude that topology should be prioritized for further development because of its potential to contribute to biology. Instead, the committee expects that biological problems will continue to drive the importation and evolution of applicable mathematics. Then, as general principles emerge, they will be codified at the appropriate level of generality. For machine learning, HMMs, and Monte Carlo sampling, this process is well under way. Indeed, these powerful methods are now well established in the toolkits of most computational biologists and are routinely taught in introductory graduate-level courses covering computational biology. Other methods will follow, just as others went before. The greatest enabler of this process will be research programs and collaborations that confront mathematical scientists with specific problems drawn from across the whole landscape of modern biology.

PROCESSING OF LOW-LEVEL DATA

The purpose of the current chapter, "Crosscutting Themes," is to call attention to issues that might have been neglected if the committee had relied entirely on levels of biological organization to structure this report. By discussing examples of mathematical themes that are important at many levels of biological organization, the committee accomplishes that purpose. Another quite different crosscutting theme is the importance of low-level data processing. Indeed, one could argue that the most indispensable applications of mathematics in biology have historically been in this area. Furthermore, the importance of low-level data processing in biology appears likely to grow. Rapid advances in technologies such as optics, digital electronics, sensors, and small-scale fabrication ensure that biologists will have access to ever more powerful instruments.

Nearly all the data that biologists obtain from these instruments has gone through extensive analog and digital transformations. Because these transformations improve signal-to-noise ratios, correlate signals with real-world landmarks, eliminate distortions, and otherwise add value to the physical output of the primary sensing devices, they are often the key to success during instrumentation development. The continued involvement of mathematicians, physicists, engineers, chemists, and bioscientists in instrumentation development has great potential to advance the biological sciences. Mathematical scientists are essential partners in these collaborations. Indeed, many of the challenges that arise in low-level data processing can only be met by applying powerful, abstract formalisms that are unfamiliar to most bioscientists. A few examples, discussed below, illustrate current research in this area.

In optical imaging, the development of two-photon (or, more generally, multiphoton) fluorescence microscopy is already having a significant impact on biology (So et al., 2000). This technique, in which molecular excitation takes place from the simultaneous absorption of two or more photons by a fluorophore, offers submicron resolution with relatively little damage to samples. The latter feature is of particular importance in biology since there is growing interest in observing living cells as they undergo complex developmental changes. The sensitivity of two-photon microscopy, in contrast to conventional fluorescence microscopy, is more dependent on peak illumination of the sample than on average illumination; hence, pulsed-laser light sources can be used to provide high instantaneous illumination while maintaining low average-power dissipation in the sample. Significant progress has been made in using two-photon methods to image cells, subcellular components, and macromolecules. Substantial improvements in sensitivity remain possible since in current instruments, only a small fraction of emitted photons reaches the detector. This low sensitivity, among other problems, limits the time resolution of two-photon microscopy. Computation and simulation will play a key role in efforts to increase sensitivity by optimizing the light path and improving detectors. Discussing the potential of future improvements in the sensitivity of two-photon microscopy, Fraser (2003) observed that "with a combined improvement of only ten-fold, today's impossible project can become tomorrow's routine research project." This rapid progression from the impossible to the routine is the story of much of modern experimental biology.

An entirely different class of imaging techniques, broadly referred to as near-field microscopy, has also made great strides in recent years. Steadily improving fiber-optic light sources and detectors have been the critical enabling technologies. Optical resolutions of 20 to 50 nm are achievable with ideal samples, dramatically breaching the wavelength limit on the resolution of traditional light microscopes. Nonetheless, near-field microscopy is difficult to apply in biology because of the irregular nature of biological materials. Despite these difficulties, Doyle et al. (2001) succeeded in imaging actin filaments in glial cells, and it is reasonable to expect further progress, based in part on improved computational techniques for extracting the desired signal from the noise in near-field data.

At still higher spatial resolution, many new techniques have been introduced for the structural analysis of biological macromolecules. Examples include high-field NMR, cryo-electron microscopy (cryo-EM) (Henderson, 2004; Carragher et al., 2004), time-resolved structural analysis based on physical and chemical trapping (Hajdu et al., 2000), small-angle scattering (Svergun et al., 2002), and total-internal-reflection fluorescence microscopy (Mashanov et al., 2003). Cryo-EM has achieved 0.4-nm resolution for two-dimensional crystals and may soon achieve that capability for single particles. One problem with all imaging methods is the lack of rigorous validation methods for determining the reliability of determined structures. Henderson (2004) emphasized this point, stating that the lack of such methods is "probably the greatest challenge facing cryo-EM." The mathematical sciences have a clear role to play in addressing this challenge.

Hyperspectral imaging is the final example here of promising technologies that could be incorporated into many types of biological instrumentation. This technology involves measuring the optical response of a sample over an entire frequency range rather than at one, or a few, selected frequencies. In hyperspectral detectors, each pixel contains a spectrum with tens to thousands of measurements and allows for far more detailed characterization of a sample than could be obtained from data collected at a single frequency. Hyperspectral imaging is already being used for microscopy (Sinclair et al., 2004; Schultz et al., 2001), pathological studies (Davis et al., 2003), and microarray analysis (Sinclair et al., 2004; Schultz et al., 2001). Sinclair et al. (2004) recently developed a scanner with high spatial resolution that records an emission spectrum for each pixel over the range 490-900 nm at 3-nm intervals. These investigators used multivariate curve-resolution algorithms to distinguish between the emission spectra of the components of multiple samples. Further mathematical developments have the potential to enhance instrument design and performance for diverse applications. Similar comments apply to many aspects of imaging technology. Indeed, the committee believes that one of the important goals of the next decade in instrumentation should be to improve the quantitation achievable in all forms of biological imaging. Nearly all applications of the mathematical sciences to biology will be promoted by improved instrumentation that lowers the cost of acquiring reliable quantitative data and increases the collection rates.

EPILOGUE

This brief discussion of the role of the mathematical sciences in the development of instrumentation is a suitable note on which to conclude this report since it emphasizes the primacy of data in the interplay between mathematics and biology. Mathematical scientists, and the funding agencies that support them, should be encouraged to take an interest in the full cycle of experimental design, data acquisition, data processing, and data interpretation through which bioscientists are expanding their understanding of the living world. Applications of the mathematical sciences to biology are not yet so specialized as to make this breadth of view impractical. An illustrative case is that of Phil Green, whose training before an early-career switch to genetics was in pure mathematics. During the Human Genome Project, he made key contributions to problems at every level of genome analysis: the phred-software package transformed large-scale DNA sequencing by attaching statistically valid quality measures to the raw base calls of automated sequencing instruments (Ewing and Green, 1998; Ewing et al., 1998); phrap, consed, and autofinish software sheperded these base calls all the way to finished-DNA sequence (Gordon et al., 1998; Gordon et al., 2001); then, in analyzing the sequence itself, Green contributed to problems as diverse as estimating the number of human genes (Ewing and Green, 2000), discovering the likely existence of a new DNA-repair process in germ cells (Green et al., 2003), and modeling sequence-context effects on mutation rates (Hwang and Green, 2004).

As this and many other stories emphasize, applications of the mathematical sciences to the biosciences span an immense conceptual range, even when one considers only one facet of the biological enterprise. No one scientist, mathematical or biological specialty, research program, or funding agency can span the entire range. Instead, the integration of diverse skills and perspectives must be the overriding goal. In this report, the committee seeks to encourage such integration by putting forward a set of broad principles that it regards as essential to the health of one of the most exciting and promising interdisciplinary frontiers in 21st century science.

REFERENCES

Adcock, C.J. 1997. Sample size determination: A review. *Statistician* 46(2): 261-283.

Alexandersson, M., S. Cawley, and L. Pachter. 2003. SLAM—Cross-species gene finding and alignment with a generalized pair hidden Markov model. *Genome Res.* 13(3): 496-502.

Alizadeh, A.A., M.B. Eisen, R.E. Davis, C. Ma, I.S. Lossos, A. Rosenwald, J.C. Boldrick, H. Sabet, T. Tran, X. Yu, J.I. Powell, L. Yang, G.E. Marti, T. Moore, J. Hudson Jr., L. Lu, D.B. Lewis, R. Tibshirani, G. Sherlock, W.C. Chan, T.C. Greiner, D.D. Weisenburger, J.O. Armitage, R. Warnke, R. Levy, W. Wilson, M.R. Grever, J.C. Byrd, D. Botstein, P.O. Brown, and L.M. Staudt. 2000. Distinct types of diffuse large B-cell lymphoma identified by gene expression profiling. *Nature* 403(6769): 503-511.

Alter, O., P.O. Brown, and D. Botstein. 2000. Singular value decomposition for genome-wide expression data processing and modeling. *Proc. Natl. Acad. Sci. U.S.A.* 97(18): 10101-10106.

Baldi, P., and A.D. Long. 2001. A Bayesian framework for the analysis of microarray expression data: Regularized t-test and statistical inferences of gene changes. *Bioinformatics* 17(6): 509-519.

Becquet, C., S. Blachon, B. Jeudy, J.-F. Boulicaut, and O. Gandrillon. 2002. Strong-association-rule mining for large-scale gene-expression data analysis: A case study on human SAGE data. *Genome Biol.* 3(12): Research0067.

Bittner, M., P. Meltzer, Y. Chen, Y. Jiang, E. Seftor, M. Hendrix, M. Radmacher, R. Simon, Z. Yakhini, A. Ben-Dor, N. Sampas, E. Dougherty, E. Wang, F. Marincola, C. Gooden, J. Lueders, A. Glatfelter, P. Pollock, J. Carpten, E. Gillanders, D. Leja, K. Dietrich, C. Beaudry, M. Berens, D. Alberts, and V. Sondak. 2000. Molecular classification of cutaneous malignant melanoma by gene expression profiling. *Nature* 406(6795): 536-540.

Brunet, J.P., P. Tamayo, T.R. Golub, and J.P. Mesirov. 2004. Metagenes and molecular pattern discovery using matrix factorization. *Proc. Natl. Acad. Sci. U.S.A.* 101(12): 4164-4169.

Burge, C., and S. Karlin. 1997. Prediction of complete gene structures in human genomic DNA. *J. Mol. Biol.* 268(1): 78-94.

Califano, A. 2000. SPLASH: Structural pattern localization analysis by sequential histograms. *Bioinformatics* 16(4): 341-357.

Carragher, B., D. Fellmann, F. Guerra, R.A. Milligan, F. Mouche, J. Pulokas, B. Sheehan, J. Quispe, C. Suloway, Y. Zhu, and C.S. Potter. 2004. Rapid, routine structure determination of macromolecular assemblies using electron microscopy: Current progress and further challenges. *J. Synchrotron Rad.* 11: 83-85.

Cho, R.J., M.J. Campbell, E.A. Winzeler, L. Steinmetz, A. Conway, L. Wodicka, T.G. Wolfsberg, A.E. Gabrielian, D. Landsman, D.J. Lockhart, and R.W. Davis. 1998. A genome-wide transcriptional analysis of the mitotic cell cycle. *Mol. Cell* 2: 65-73.

Churchill, G.A. 1989. Stochastic models for heterogeneous DNA sequences. *Bull. Math. Bio.* 51(1): 79-94.

Cortes, C., L.D. Jackel, and W.-P. Chiang. 1995. Limits on learning machine accuracy imposed by data quality. Pp. 57-62 in *Proceedings of the First International Conference on Knowledge Discovery and Data Mining*. U.M. Fayyad and R. Uthurusamy, eds. Montreal, Canada: AAAI Press.

Cortes, C., L.D. Jackel, S.A. Solla, V. Vapnik, and J.S. Denker. 1993. Learning curves: Asymptotic values and rate of convergence. Pp. 327-334 in *Advances in Neural Information Processing Systems*. NIPS'1993, Vol. 6. Denver, Colo.: Morgan Kauffman.

Davis, G.L., M. Maggioni, R.R. Coifman, D.L. Rimm, and R.M. Levenson. 2003. Spectral/spatial analysis of colon carcinoma. *Modern Pathol.* 16 (1): 320A-321A.

Doyle, R.T., M.J. Szulzcewski, and P.G. Haydon. 2001. Extraction of near-field fluorescence from composite signals to provide high resolution images of glial cells. *Biophys. J.* 80: 2477-2482.

Duda, R.O., P.E. Hart, and D.G. Stork. 2000. *Pattern Classification*. New York, N.Y.: John Wiley & Sons Ltd.

Dudoit, S., Y.H. Yang, M.J. Callow, and T.P. Speed. 2002. Statistical methods for identifying differentially expressed genes in replicated cDNA microarray experiments. *Statistica Sinica* 12(1): 111-139.

Eisen, M.B., P.T. Spellman, P. Brown, and D. Botstein. 1998. Cluster analysis and display of genome-wide expression patterns. *Proc. Natl. Acad. Sci. U.S.A.* 95(25): 14863-14868.

Evgeniou, T., M. Pontil, and T. Poggio. 2000. Regularization networks and support vector machines. *Adv. Comput. Math.* 13: 1-50.

Ewing, B., and Green P. 1998. Base-calling of automated sequencer traces using phred. II. Error probabilities. *Genome Res.* 8(3): 186-194.

Ewing, B., and Green P. 2000. Analysis of expressed sequence tags indicates 35,000 human genes. *Nat. Genet.* 25(2): 232-234.

Ewing, B., L. Hillier, M.C. Wendl, and P. Green. 1998. Base-calling of automated sequencer traces using phred. I. Accuracy assessment. *Genome Res.* 8(3): 175-185.

Fraser, S.E. 2003. Crystal gazing in optical microscopy. *Nat. Biotechnol.* 21(11): 1272-1273.

Frenkel, D., and B. Smit. 1996. *Understanding Molecular Simulation: From Algorithms to Applications*. San Diego, Calif.: Academic Press.

Friedman, J.H. 1994. An overview of computational learning and function approximation. Pp. 1-61 in *From Statistics to Neural Networks. Theory and Pattern Recognition Applications*. V. Cherkassky, J.H. Friedman, and H. Wechsler, eds. Berlin: Springer-Verlag.

Friedman, N., M. Linial, I. Nachman, and D. Pe'er. 2000. Using Bayesian networks to analyze expression data. *J. Comput. Biol.* 7: 601-620.

Gelfand, A.E., and A.F.M. Smith. 1990. Sampling-based approaches to calculating marginal densities. *J. Am. Stat. Assoc.* 85: 398-409.

Geman, S., and D. Geman. 1984. Stochastic relaxation, Gibbs distributions and the Bayesian restoration of images. *IEEE T. Pattern Anal.* 6: 721-741.

Gilks, W.R., S. Richardson, and D.J. Spegelhalter. 1996. *Markov Chain Monte Carlo in Practice*. London, England: Chapman and Hall.

Golub, T.R., D.K. Slonim, P. Tamayo, C. Huard, M. Gaasenbeek, J.P. Mesirov, H. Coller, M.L. Loh, J.R. Downing, M.A. Caligiuri, C.D. Bloomfield, and E.S. Lander. 1999. Molecular classification of cancer: Class discovery and class prediction by gene expression monitoring. *Science* 286(5439): 531-537.

Gordon, D., C. Abajian, and P. Green. 1998. Consed: A graphical tool for sequence finishing. *Genome Res.* 8(3): 195-202.

Gordon, D., C. Desmarais, P. Green. 2001. Automated finishing with autofinish. *Genome Res.* 11(4): 614-625.

Green, P., B. Ewing, W. Miller, P.J. Thomas, and E.D. Green. 2003. Transcription-associated mutational asymmetry in mammalian evolution. *Nat. Genet.* 33(4): 514-517.

Gribskov, M., A.D. McLachlan, and D. Eisenberg. 1987. Profile analysis: Detection of distantly related proteins. *Proc. Natl. Acad. Sci. U.S.A.* 84(13): 4355-4358.

Hajdu, J., R. Neutze, T. Sjögren, K. Edman, A. Szöke, R.C. Wilmouth, and C.M. Wilmot. 2000. Analyzing protein functions in four dimensions. *Nat. Struct. Biol.* 7(11): 1006-1012.

Hansmann, U.H.E., M. Masuya, and Y. Okamoto. 1997. Characteristic temperatures of folding of a small peptide. *Proc. Natl. Acad. Sci. U.S.A.* 94: 10652-10656.

Hao, M.-H., and H.A. Scheraga. 1998. Molecular mechanisms of coperative folding of proteins. *J. Mol. Biol.* 277: 973-983.

Hastie, T., R. Tibshirani, and J. Friedman. 2001. *The Elements of Statistical Learning.* New York, N.Y.: Springer.

Henderson, R. 2004. Realizing the potential of electron cryo-microscopy. *Q. Rev. Biophys.* 37(1): 3-13.

Hwang, D.G., and P. Green. 2004. Bayesian Markov chain Monte Carlo sequence analysis reveals varying neutral substitution patterns in mammalian evolution. *Proc. Natl. Acad. Sci. U.S.A.* 101(39): 13994-14001.

Ideker, T., V. Thorsson, A.F. Siegel, and L.E. Hood. 2000. Testing for differentially-expressed genes by maximum-likelihood analysis of microarray data. *J. Comput. Biol.* 7(6): 805-817.

Kim, P.M., and B. Tidor. 2003. Subsystem identification through dimensionality reduction of large-scale gene expression data. *Genome Res.* 13(7): 1706-1718.

Kluger, Y., R. Basri, J.T. Chang, and M. Gerstein. 2003. Spectral biclustering of microarray data: Coclustering genes and conditions. *Genome Res.* 13(4): 703-716.

Korf, I., P. Flicek, D. Duan, and M.R. Brent. 2001. Integrating genomic homology into gene structure prediction. *Bioinformatics* 17(Suppl 1): S140-S148.

Krogh, A., M. Brown, I.S. Mian, K. Sjolander, and D. Haussler. 1994. Hidden Markov models in computational biology: Applications to protein modeling. *J. Mol. Biol.* 235(5): 1501-1531.

Kulp, D., D. Haussler, M.G. Reese, and F.H. Eeckman. 1996. A generalized Hidden Markov Model for the recognition of human genes in DNA. *Proc. Int. Conf. Intell. Syst. Mol. Biol.* 4: 134-142.

Lauritzen, S.L., and D.J. Speigelhalter. 1988. Local computations with probabilities on graphical structures and their application to expert systems. *J. Roy. Stat. Soc. B* 50: 157-224.

Lawrence, C.E., S.F. Altschul, M.S. Boguski, A.F. Neuwald, and J.C. Wooton. 1993. Detecting subtle sequence signals: A Gibbs sampling strategy for multiple alignment. *Science* 262: 208-214.

Lazzeroni, L., and A.B. Owen. 2002. Plaid models for gene expression data. *Stat. Sinica* 12(1): 61-86.

Lee, D.D., and H.S. Seung. 1999. Learning the parts of objects by non-negative matrix factorization. *Nature* 401(6755): 788-791.

Lee, M.L., F.C. Kuo, G.A. Whitmore, and J. Sklar. 2000. Importance of replication in microarray gene expression studies: Statistical methods and evidence from repetitive cDNA hybridizations. *Proc. Natl. Acad. Sci. U.S.A.* 97(18): 9834-9839.

Liu, J.S. 2001. *Monte Carlo Strategies in Scientific Computing.* New York, N.Y.: Springer-Verlag.

Mashanov, G.I., D. Tacon, A.E. Knight, M. Peckham, and J.E. Molloy. 2003. Visualizing single molecules inside living cells using total internal reflection fluorescence microscopy. *Methods* 29: 142-152.

Metropolis, N., A.W. Rosenbluth, M.N. Rosenbluth, A.H. Teller, and E. Teller. 1953. Equations of state calculations by fast computing machines. *J. Chem. Phys.* 21: 1087-1091.

Meyer, I.M., and R. Durbin. 2002. Comparative ab initio prediction of gene structures using pair HMMs. *Bioinformatics* (10): 1309-1318.

Minsky, M., and S. Papert. 1988. *Perceptrons. An Introduction to Computational Geometry.* Cambridge, Mass.: MIT Press.

Mootha, V.K., C.M. Lindgren, K.F. Eriksson, A. Subramanian, S. Sihag, J. Lehar, P. Puigserver, E. Carlsson, M. Ridderstrale, E. Laurila, N. Houstis, M.J. Daly, N. Patterson, J.P. Mesirov, T.R. Golub, P. Tamayo, B. Spiegelman, E.S. Lander, J.N. Hirschhorn, D. Altshuler, and L.C. Groop. 2003. PGC-1 alpha-responsive genes involved in oxidative phosphorylation are coordinately downregulated in human diabetes. *Nat. Genet.* 34(3): 267-273.

Mukherjee, S., P. Tamayo, S. Rogers, R. Rifkin, A. Engle, C. Campbell, T.R. Golub, and J.P. Mesirov. 2003. Estimating dataset size requirements for classifying DNA microarray data. *J. Comput. Biol.* 10(2): 119-142.

Murali, T.M., and S. Kasif. 2003. Extracting conserved gene expression motifs from gene expression data. Pp. 77-88 in *Pacific Symposium on Biocomputing 2003*. Singapore: World Scientific.

Nature Genetics Supplement 21. 1999.

Nature Genetics Supplement 32. 2002.

Newton, M.A., C.M. Kendziorski, C.S. Richmond, F.R. Blattner, and K.W. Tsui. 2001. On differential variability of expression ratios: Improving statistical inference about gene expression changes from microarray data. *J. Comput. Biol.* 8(1): 37-52.

Perou, C.M., S.S. Jeffrey, M. van de Rijn, C.A. Rees, M.B. Eisen, D.T. Ross, A. Pergamenschikov, C.F. Williams, S.X. Zhu, J.C. Lee, D. Lashkari, D. Shalon, P.O. Brown, and D. Botstein. 1999. Distinctive gene expression patterns in human mammary epithelial cells and breast cancers. *Proc. Natl. Acad. Sci. U.S.A.* 96(16): 9212-9217.

Perou, C.M., T. Sorlie, M.B. Eisen, M. van de Rijn, S.S. Jeffrey, C.A. Rees, J.R. Pollack, D.T. Ross, H. Johnsen, L.A. Akslen, O. Fluge, A. Pergamenschikov, C. Williams, S.X. Zhu, P.E. Lonning, A.L. Borresen-Dale, P.O. Brown, and D. Botstein. 2000. Molecular portraits of human breast tumours. *Nature* 406(6797): 747-752.

Pomeroy, S.L., P. Tamayo, M. Gaasenbeek, L.M. Sturla, M. Angelo, M.E. McLaughlin, J.Y. Kim, L.C. Goumnerova, P.M. Black, C. Lau, J.C. Allen, D. Zagzag, J.M. Olson, T. Curran, C. Wetmore, J.A. Biegel, T. Poggio, S. Mukherjee, R. Rifkin, A. Califano, G. Stolovitzky, D.N. Louis, J.P. Mesirov, E.S. Lander, and T.R. Golub. 2002. Prediction of central nervous system embryonal tumour outcome based on gene expression. *Nature* 415(6870): 436-442.

Rosenblatt, F. 1962. *Principles of Neurodynamics.* New York, N.Y.: Spartan Books.

Roweis, S.T., and L.K. Saul. 2000. Nonlinear dimensionality reduction by locally linear embedding. *Science* 290(5500): 2323-2326.

Schultz, R.A., T. Nielsen, J.R. Zavaleta, R. Ruch, R. Wyatt, and H.R.Garner. 2001. Hyperspectral imaging: A novel approach for microscopic analysis. *Cytometry* 43(4): 239-247.

Searls, D.B. 1992. The linguistics of DNA. *Am. Sci.* 80: 579-591.

Sela, M., F.H. White Jr., and C.B. Anfinsen. 1957. Reductive cleavage of disulfide bridges in ribonuclease. *Science* 125: 691-692.

Shipp, M.A., K.N. Ross, P. Tamayo, A.P. Weng, J.L. Kutok, R.C. Aguiar, M. Gaasenbeek, M. Angelo, M. Reich, G.S. Pinkus, T.S. Ray, M.A. Koval, K.W. Last, A. Norton, T.A. Lister, J. Mesirov, D.S. Neuberg, E.S. Lander, J.C. Aster, and T.R. Golub. 2002. Diffuse large B-cell lymphoma outcome prediction by gene-expression profiling and supervised machine learning. *Nat. Med.* 8(1): 68-74.

Sinclair, M.B., J.A. Timlin, D.M. Haaland, and M. Werner-Washburne. 2004. Design, construction, characterization, and application of a hyperspectral microarray scanner. *Appl. Optics* 43 (10): 2079-2088.

Slonim, D., P. Tamayo, J.P. Mesirov, T.R. Golub, and E.S. Lander. 2000. Class prediction and discovery using gene expression data. Pp. 263-272 in *Proceedings of Fourth Annual International Conference on Computational Molecular Biology.* New York, N.Y.: ACM Press.

So, P.T.C., C.Y. Dong, B.R. Masters, and K.M. Berland. 2000. Two-photon excitation fluorescence microscopy. *Ann. Rev. Biomed. Eng.* 2: 399-429.

Staunton, J.E., D.K. Slonim, H.A. Coller, P. Tamayo, M.J. Angelo, J. Park, U. Scherf, J.K. Lee, W.O. Reinhold, J.N. Weinstein, J.P. Mesirov, E.S. Lander, and T.R. Golub. 2001. Chemosensitivity prediction by transcriptional profiling. *Proc. Natl. Acad. Sci. U.S.A.* 98(19): 10787-10792.

Stormo, G.D., and D. Haussler. 1994. Optimally pairing a sequence into different classes based on multiple types of evidence. Pp. 369-375 in *Proceedings of the Second International Conference on Intelligent Systems for Molecular Biology.* Vol. 2. R. Altman, D. Brutlag, P. Karp, R. Lathrop, and D. Searls, eds. Menlo Park, Calif.: AAAI Press.

Sumners, D. 1995. Lifting the curtain: Using topology to probe the hidden action of enzymes. *Notices of the AMS* 42: 528-537.

Svergun, D.I., and M.H.J. Koch. 2002. Advances in structure analysis using small-angle scattering in solution. *Curr. Opin. Struct. Biol.* 12: 654-660.

Tamayo, P., D. Slonim, J. Mesirov, Q. Zhu, S. Kitareewan, E. Dmitrovsky, E.S. Lander, and T.R. Golub. 1999. Interpreting patterns of gene expression with self-organizing maps: Methods and application to hematopoietic differentiation. *Proc. Natl. Acad. Sci. U.S.A.* 96(6): 2907-2912.

Tanner, M.A. 1996. *Tools for Statistical Inference: Methods for the Exploration of Posterior Distributions,* 3rd ed. New York, N.Y.: Springer-Verlag.

Tanner, M.A., and W.H. Wong. 1987. The calculation of posterior distributions by data augmentation (with discussion). *J. Am. Stat. Assoc.* 82: 528-550.

Tenenbaum, J.B., V. de Silva, and J.C. Langford. 2000. A global geometric framework for nonlinear dimensionality reduction. *Science* 290(5500): 2319-2323.

Tusher, V.G., R. Tibshirani, and G. Chu. 2001. Significance analysis of microarrays applied to the ionizing radiation response. *Proc. Natl. Acad. Sci. U.S.A.* 98(9): 5116-5121.

Vapnik, V. 1998. *Statistical Learning Theory.* New York, N.Y.: John Wiley & Sons Ltd.

von Dassow, G., E. Meir, E.M. Munro, and G.M. Odell. 2000. The segment polarity network is a robust developmental module. *Nature* 406: 188-192.

Winfree, A.T. 1983. Sudden cardiac death, a problem in topology. *Sci. Am.* 248: 114-161.